D0847365

DATE DUE

1981

ORIGIN OF
GRANITE BATHOLITHS
Geochemical Evidence

ORIGIN OF GRANITE BATHOLITHS
Geochemical Evidence

Based on a meeting of the Geochemistry Group of the Mineralogical Society

M. P. Atherton and J. Tarney (Editors)
University of Liverpool and University of Birmingham

Shiva Publishing Limited

SHIVA PUBLISHING LIMITED
9 Clareville Road, Orpington, Kent BR5 1RU, UK

British Library Cataloguing in Publication Data

Origin of granite batholiths.
1. Granite 2. Batholiths
I. Atherton, M P   II. Tarney, J
III. Mineralogical Society. *Geochemistry Group*
552'.3   QE462.G7

ISBN 0–906812–00–3

Typeset and printed by Imprint (Print & Design) Ltd, Exeter, Devon

CONTENTS

PREFACE

This book is for undergraduates, postgraduates and research workers who wish to gain an insight into present ideas and speculations on the origin of granite batholiths. It is a summary of the proceedings of a one-day meeting of the Geochemistry Group of the Mineralogical Society held at the University of Liverpool on the 2nd May 1979, entitled *The Origin of Granite Batholiths: Geochemical Evidence.* It was felt that relevant new geochemical and isotopic data with associated field and petrological observations would be helpful in clarifying the main issues connected with the origin of granitic rocks. The speakers who participated contributed a wealth of data and ideas to the problem, based on many years' experience of granitic rocks in various parts of the world. The popularity of the meeting indicated that a summary of the talks might be welcomed by a wider audience; hence this book. For reasons of length and cost, authors have been limited to presenting only the essential results necessary to develop their arguments and ideas. The comprehensive reference list will, however, assist those who wish to pursue specific aspects in greater detail. Inevitably, the cost and speed of publication may have resulted in some errors and inconsistencies which would not have occurred in a book produced over a longer time-span, but the editors hope that the rapid and up-to-date publication will offset this.

The controversy over the origin of granite is as old as the science of geology. There can be few subjects in the science which have taken up so much journal space over the years, but where a consensus solution is still awaited – a surprising situation, perhaps, for a rock type which is mineralogically simple. Yet the problem is as fundamental as the origin of the continental crust itself.

Experimental studies over the last few decades have successfully elucidated the possible ways in which granitic magmas can be produced: by partial melting of various source rocks or by fractional crystallization. Geochemical data can, however, provide valuable constraints on the various probable mechanisms. Moreover, it is clear from this volume that despite the undoubted progress which has been made, the problem of the origin of granitic batholiths is still with us. The perceptive writings of H.H. Read thirty years ago have conditioned us into accepting that there are 'granites and granites'. Perhaps, in view of the geochemical evidence presented here, it would be more appropriate to admit that there are 'granites, granites and still more granites'. It is also clear that for a proper understanding of the petrogenesis and geochemical history of a particular batholith it is essential that detailed field, structural and petrographic studies must be made before sampling is attempted – only then will the constraints be properly realized. We would particularly like to thank the authors, who all managed to supply their final typescripts so promptly after the meeting, and L.S. Sanderson, M.S. Skwarnecki and A.D. Saunders who assisted in various ways in the preparation of the book. M.P. Atherton would also like to thank W.S. Pitcher, J. Cobbing, N. Moore, L. Aguirre, C. Vidal and M.A. Bussell for much instruction and insight into the complexities and beauty of batholithic rocks and their associated volcanics.

M.P. ATHERTON, Liverpool
October 1979   J. TARNEY, Birmingham

COMMENTS ON THE GEOLOGICAL ENVIRONMENTS OF GRANITES

WALLACE S. PITCHER
The Jane Herdman Laboratories of Geology, University of Liverpool,
Brownlow Street, P.O. Box 147, Liverpool L69 3BX, UK

Field studies of granitic batholiths convincingly indicate that there is so long and complex a history of emplacement, involving compositional variations and even mixtures of genetic types (e.g. Pitcher and Berger, 1972; Pitcher, 1978, 1979), that simple reconnaissance geochemical studies will not suffice to unravel the evolution and genesis of the constitutent magmas. Clearly, the establishment of a geological space-time framework is a necessary prerequisite to such studies.

At the onset it is important to appreciate that the numerous individual plutons which constitute a gregarious batholith, each one having its own evolutionary history, are yet derived from a limited number of unique magma types. These are often quite separate in time of generation, and appear in the crust within many separate plutons distributed throughout a batholith (Cobbing, Pitcher and Taylor, 1977). Such magma types are represented in the *super-units* of Cobbing and Pitcher (1972), the *magma sequences* of Bateman and Dodge (1970), the *suites* of White, Williams and Chappell (1977) and, possibly, the *isotopic domains* of Silver (personal communication). As an example the Lima segment of the Coastal Batholith of Peru comprises seven such sequences, the Arequipa segment four, and these are distributed within some 700–800 separate plutons.

It is even more important to realize that there is no one origin of granite – according to Read (1948 in 1957) there are granites and granites – so that there can be no single path of evolution of the magmas. Thus, in the anorogenic context granitoids can be produced either by the differentiation of basic magma or by the partial melting of crustal material by basic magma, while in the orogenic context they may be generated either by complex partial remelting of mantle or lower crust, or as the end-stage of metamorphism. The resulting genetic types have been separately designated, particularly by Russian authors (e.g. Tauson and Kozlov, 1973) who are foremost both in understanding the economic importance of Read's dictum and in describing the chemical characteristics of each metallogenic granite type (see Levinson, 1974, for details).

Although such a genetic classification is in its infancy, it is quite clear that the geochemical characteristics of the several granitic associations will be very different. Thus, the metallogenetically barren, granophyric granites that are associated with gabbros, such as occur in the Tertiary volcanic centres of the British Isles, contrast strongly with the niobium-rich, mildly alkaline granites, such as characterize the Mesozoic centred complexes of Nigeria, and the distinctly alkaline, 'agpaitic' syenogranites, of which those of Permian age around Oslo are so excellent

1

an example. Different again are the often tin-bearing, two-mica granites which dominate the European Hercynian, and the calc-alkaline granitoids which are so abundant in the Mesozoic Andean and which have an important porphyry-copper association. Such differences are most clearly related to their constrasted geological environments and so, ultimately, to their separate origins.

There is, however, an important generality concerning granites whatever the environment. While scientific fashion dictates the consensus opinion of the time concerning the various origins, it seems undeniable that remelting and metasomatic processes require so considerable an energy input that it is most unlikely that granites in any quantity can be produced *in situ* out of harmony in time and place with their metamorphic environments (Walton, 1955). Very many granitic plutons show an energy disharmony with their country rocks so that they are likely to have been intruded as magmas, whatever their origin at depth.

A further common denominator in this discussion is that granites occur predominantly in association with continental crust, and in batholithic volumes only within mobile belts marginal to or within such a crust. This fact has been long understood in terms of a local thickening of the crust providing a sufficient rise in temperature to produce crustal melts (see Daly, 1933, p. 263), although this has now been restated in terms of magma production at plate boundaries, where crustal thickening takes place as a consequence of underplating (by subduction) or crustal-shortening (by collision).

It has also long been recognized that there is a general space-time connection between regional deformation, metamorphism and the production and emplacement of granite. This has been formalized in the magma-tectonic grouping of Eskola (1932) and Wegmann (1935), and elaborated by Read (1957), Buddington (1959), Hutchinson (1970) and Stephannson (1975), the latter introducing (see his Table 1, p. 193) yet another categorization, *viz.* epitectonic, mesotectonic and catatectonic plutons. Such groupings, which are more familiarly referred to in terms of epizonal, mesozonal and catazonal, must reflect the ductility contrast between the intrusive body and the country rocks, but this is itself a reflection of three major variables, *viz.* the lithological contrast, the relative crustal level of emplacement, and the time relationship between intrusion and regional metamorphism.

One implication of such connections is that granite can be a natural end-stage product of prograde metamorphism. However, the relationship is not necessarily a direct one if only because the intrusion of granite in great volume is often considerably divorced in time from the apogee of metamorphism — the end-Silurian granites of Caledonia are a prime example. It is the thickening of the crust which provides the common factor, and this is accomplished in different ways, e.g. by underplating in a subduction regime, and by crustal shortening in a collision regime.

Furthermore, it is a paradox of our times that although experimental work has conclusively proved that granitoids could be produced by the partial melting of continental rocks (Winkler, 1976; Wyllie, 1977), yet we are being increasingly reminded that neither the temperatures nor the water-concentrations during ultrametamorphism in the deep crust are sufficiently high to permit remelting to

2

occur (e.g. Brown and Hennessy, 1978), at least not without supplementation via the agency of mantle-derived magmas (see Brown, this volume). But in this respect, as in much else, there is a great difference between granites generated in collision-type and subduction-type mobile belts (see Beckinsale, this volume). In the latter, remelting of the dehydrated, underplated crust is unlikely to occur without this introduction of heat and volatiles from elsewhere, but in collision-type belts sufficient water can surely be made available by the dehydration of the micas, particularly during prograde metamorphism of the thickening metasedimentary wedges.

However, it is not the purpose of this introductory note to embark on a discussion of the origin of granite magma, but to state a prejudice that, as noted above, we need to move towards an environmental classification of granitoids, even towards the introduction of a new nomenclature. As an example, two main types of contrasted granite series can reasonably be identified, even though there are clearly intermediate types (see Coleman, this volume), which appear to characterize two different types of orogenic belt. Both series are calc-alkaline but one is compositionally expanded, the other compositionally restricted, a difference most clearly expressed in terms of the gabbro-diorite/tonalite-granodiorite/granite proportion: i.e. the former is typified by a proportion near 15:50:35, the other by one near 2:18:80 (Pitcher, 1979).

Again, this is no new finding (see Eskola, 1932), but in recent years it has been seen to be reflected by marked differences in the various geochemical parameters, particularly in those measured by Chappell and White (1974; see also Hine *et al.*, 1978), and which involve the specific recognition of granitoids of a crustal, S-type, generally compositionally restricted, and a mantle, I-type, commonly compositionally expanded, in the terms introduced above. Other parameters follow suit, such as the presence or absence of muscovite, the values of the initial $^{87}Sr/^{86}Sr$ ratios, the latter being relatively higher for S-types, with that of 0.7060 seeming to provide a critical boundary (e.g. Kistler, 1974; Armstrong *et al.*, 1977); also the presence or absence of inherited xenocrysts of crustal zircons (Silver and Deutsch, 1963; and see Pidgeon and Aftalion, 1978). Such differences can be reflected in the normative corundum, or the $^{16}O/^{18}O$ ratio (Taylor, 1977; O'Neil *et al.*, 1977), or the relative importance of magnetite to ilmenite (Ishihara, 1977), or the sphene–allanite and ilmenite–monazite pairs (Ivanova and Butuzova, 1968) — the latter two comparatives specifically reflecting the different oxygen fugacities in I- and S-types, respectively.

It is not surprising that the composition of granitoid magmas should reflect differences in source. I-type granitoids, derived from a basic igneous source, will differ from S-types, derived from a metasedimentary source, by greater abundance of the basic cognates, by the greater regularity in compositional variation, by the 'igneous' derivation of the restitic material, by a different chemical relation between melt and restite, and also by differences in the K, Rb and Sr proportions and in oxidation as is appropriate to a source which has or has not been through a weathering cycle (White and Chappell, 1977). The likelihood is, however, that the provenance will not be so exclusive as to deny mixing processes, so resulting

3

in magmas with a spectrum of I- and S-characteristics (see Stephens and Halliday, this volume).

An example of a total study of magma provenance is that carried out by White, Chappell and their co-workers (1974, 1978; Hine *et al.*, 1978) in the Lacklan zone of eastern Australia. These authors recognize that a group of early, metamorphically harmonious plutons are largely composed of S-type granites and that these probably originated by the remelting of metasediments, while a later group of metamorphically disharmonious plutons (i.e. with superposed aureoles) are mainly I-types derived by remelting of deep-seated igneous material (full discussion in White and Chappell, 1977). The different types are distributed in separate, though adjacent, belts between which the dividing S–I line seems to represent a change in crustal composition (Chappell, 1978, p. 282). Of particular interest is the finding that a tin mineralization is attached to the S-type, and a porphyry-copper type mineralization to the I-type plutons.

However, we must not be so naive as to expect the recognition of such types to provide so specific an identification of the source rock, if only because I-type characteristics, for example, could equally well be attained by derivation from mantle or subducted oceanic basalt, or from composite remelts temporarily staged by subcrustal accretion, or from a crust built from volcanic rocks originally derived from these various sources. Indeed, I believe that a multicyclic origin of granitoid magmas will be found to be necessary to model the somewhat complex geochemical patterns which are emerging.

From such regional provenance studies two points emerge particularly clearly. Shilo and Milov (1977, p. 122) make the first of these in their review of granite emplacement in the north-eastern USSR, i.e. since, in that region, the depth at which plutons are revealed is of the same order in the different belts, it seems valid to hold that the petrological peculiarities of granite magmatism in these belts were determined at the level of melt generation, especially whether this lay in the crust or mantle (see also Gastil, 1975). A second general point is made by Kistler (1974, p. 413) in his concern with the Sierra Nevada, *viz.* that the regional compositional changes across the batholith are independent of the age of the granitoids. Again it is the nature (and depth) of the source which determines granite type and this is now a strongly held view advocated, for example, to explain the S-type geochemistry of the granitic rocks of Oregon, Washington and Idaho (Amstrong *et al.*, 1977).

Reference to the K-h (depth) relationship (Dickinson, 1975), especially as presently elaborated to include other geochemical parameters, introduces a note of caution. If a steady progression across batholithic zones at consuming plate-boundaries is the rule, it would be remarkable if this were to result from magma generation in source-rock end-members as contrasted as mantle and crust, although perhaps less so if the source were to be subcrustal material accreted during an earlier cycle.

Broadening the topic we find that the general interrelationship between magmatism, tectonism and crustal site has been discussed by many authors, particularly in reference to the all-important metamorphic association referred to

4

above (Read, 1951; Miyashiro, 1961, 1967; Zwart, 1967, 1969). Thus, Table 1 emphasizes the different roles of granitoids within the contrasted types of orogens and in it Zwart's (1967) comparison of the characters of the 'Hercynotype' and 'Alpinotype' orogenies has been extended to include the 'Andinotype'. Undoubtedly such tabulations are too simple if only because each mobile belt is to some degree unique, but it does seem that in ocean-continent plate-edge orogens, those involving external, volcanogenic troughs, floored by oceanic crust and undergoing high pressure–low temperature metamorphism, rarely develop granite magmas; while those involving internal, volcanogenic troughs, overlying continental crust and undergoing non-deformative depth metamorphism, support immense

Table 1 **Contrasting characters of orogens (on a structural, not time-stratigraphic basis, partly after Zwart, 1967)**

Alpinotype Orogenies:	Island-arc derived volcaniclastic sediments and lavas deposited in oceanic trenches; crustal shortening involving thrusting with nappes predominant; high pressure regional metamorphism with wide progressive zonation; ultrabasic rocks abundant. Granitoid batholiths characteristically absent
Andinotype Orogenies:	Island-arc volcaniclastic sediments and lavas deposited in troughs of eugeosynclinal type located within the continental lip and paired with belts of shelf-facies clastics; little crustal shortening but vertical movements dominant, with open, drape-folding lacking cleavage; regional burial metamorphism. Compositionally expanded, disharmonious, granitoid batholiths with important basic plutonic and andesitic volcanic associations. I-type granites with crustal involvement only in the later stages of evolution: $^{87}Sr/^{86}Sr \leqslant 0.706$; include restite material from subcrustal source
Hercynotype Orogenies:	Non-volcanic, continentally-derived sediments in intracratonic basins of miogeosynclinal type; crustal shortening with upright folding and cleavage; low-pressure metamorphism with prograde zonation; ultrabasic rocks rare. Compositionally contracted, harmonious, granitoid batholiths with only minor basic association and generally lacking contemporaneous volcanics. S-type or mixed S- and I-type granites; $^{87}Sr/^{86}Sr > 0.706$; inherited xenocrysts derived from recycled crustal rocks

multiple batholiths filled with a compositionally-expanded series of I-type granitoids, usually assembled over long periods (e.g. Pitcher, 1978). Intraplate orogens, which are widely thought to be due to continental-plate collision and which support low-pressure—low- to high-temperature metamorphism, also support batholiths, although of lesser volume and assembled over shorter periods, and compositionally dominated (but not exclusively so) by S-types with their restricted overall variation. Such contrasts are likely to be the consequence of the different mechanisms of crustal deformation because these determine the duration of the processes, the degree of participation of mantle-derived basic rocks, and even the nature of lower crust. In Hercvnotype orogens the latter is likely to be of sialic character, and be derived by tectonic thickening, while in Andinotype orogens it is likely to be a result of underplating by igneous material melted out of the oceanic crust and mantle.

It is only in the Andinotype context that there is a clear space-time relationship between plutonism and volcanism. Even here there is still uncertainty as to how direct this is — whether the plutons themselves vent any substantial volume of the volcanic material via the subvolcanic, central complexes which undoubtedly exist in the batholith environment (e.g. Bussell *et al.*, 1976; and see Atherton *et al.*, this volume), or whether the connection is simply in having a common source at depth, with separate supply routes, and with the volcanic magmas venting via fissures. My prejudice is for a dual mechanism whereby the silicic ignimbrites are vented from pluton-based calderas, the andesitic flows from fissures.

This difference between batholiths in Andinotype and Hercynotype belts extends not only to the volcanic association, the composition and genesis, but probably also to the mode of emplacement. The nature of the crust may determine whether the magmas fill predominantly cauldron or predominantly diapir batholiths (Pitcher, 1978, 1979) — whether it is the largely tensional regime of the edge of a rigid continental plate overlying a subduction zone, as in the central Andes, or the compressive, ductile regime of an intracontinental orogen, such as that of the Hercynian of south-western Europe.

To rise at all, the magmas filling the cauldrons, especially the relatively dry magmas of destructive plate margins, need to be channelled and relatively rapidly intruded along major fractures, creating a narrow heat plume within a cool upper crust. Such hot magmas, derived at least in part from the mantle, might have a sufficiently low viscosity, when isolated in their chambers, to permit differentiation *in situ* and so provide volcanic derivatives.

Magmas rising into the broad heat plume of a warm, ductile crust would be likely to form globular bodies which can diapirically expand within their ductile envelopes. The life of single diapirs might well be longer than that of the cauldrons, and the magmas, derived from the crust itself, more mushy and viscous, so that extrusion is effectively prevented.

In either case I do not envisage the plutons, in themselves, rising far in the crust; rather, that the cauldrons are filled by pumping into rock cylinders, while diapirs are filled out by pumping into expanding globules, but neither mechanism is exclusive.

6

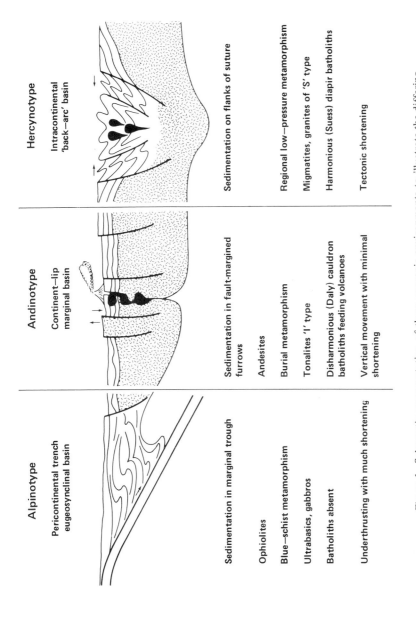

Alpinotype

Pericontinental trench
eugeosynclinal basin

Sedimentation in marginal trough

Ophiolites

Blue–schist metamorphism

Ultrabasics, gabbros

Batholiths absent

Underthrusting with much shortening

Andinotype

Continent–lip
marginal basin

Sedimentation in fault-margined
furrows

Andesites

Burial metamorphism

Tonalites 'I' type

Disharmonious (Daly) cauldron
batholiths feeding volcanoes

Vertical movement with minimal
shortening

Hercynotype

Intracontinental
'back–arc' basin

Sedimentation on flanks of suture

Regional low–pressure metamorphism

Migmatites, granites of 'S' type

Harmonious (Suess) diapir batholiths

Tectonic shortening

Figure 1 Schematic representation of three orogenic environments to illustrate the differing environments of granitic rocks

Finally, we note that the mode of intrusion, orogenic location and magma provenance are all interconnected to some degree so that we can construct a general model (Figure 1). Such model situations can clearly be assembled in nature in different ways — as coeval adjacent belts, or as sequential superimposed belts whereby, as in central Chile (Aguirre and Levi, 1977), a paired association of Alpinotype and Hercynotype belts become accreted to a plate edge, to form there the new crustal basement to a volcanic arc. It is just possible that the former, coeval situation, might model Baja California in Mesozoic times.

To conclude in most general terms, the low heat flow to be expected in an Alpinotype marginal, 'trench' environment cannot support melting at any level. In contrast, beneath a continental margin of Andinotype, subduction energetics provide sufficient heat and water to trigger remelting at various subcrustal and lower crustal levels. By such means material continuously accreting beneath the continental edge is episodically remobilized — during periods of extra rapid subduction — by hot basic magmas that melt their way up into the crust along deep-reaching crustal fractures. The new I-type magmas are permissively accepted by the rigid crust at higher levels in the form of cauldrons, often accreting to form great gregarious batholiths. Further inside the continent, within a Hercynotype fold belt, resulting perhaps by closing-collision in a back-arc situation, shortening of the crust leads to tectonic thickening when, particularly as a consequence of uplift, magmas are generated within the regionally-heated root. It is in this context that crustal-derived, S-type magmas are most likely to occur, less hot than those in the subduction regime, and be emplaced by diapiric intrusion into a ductile crust.

I have discussed just two of the possible granitic associations to illustrate that geochemical investigations must be pursued in terms of environmental models, hopefully to interact with the latter to produce more realistic hypotheses. In the event, these may show that each environment is to some extent unique — a final validation of Read's dictum concerning granites and granites.

ACKNOWLEDGMENT

The above represents a theme taken from a Presidential Address to the Geological Society of London in 1979 entitled, 'A commentary on the nature, ascent and emplacement of granitic magmas'.

8

COMPOSITIONAL VARIATION IN THE GALLOWAY PLUTONS

W.E. STEPHENS[1] and A.N. HALLIDAY[2]
[1] Department of Geology, University of St. Andrews, Fife, Scotland
[2] Scottish Universities Research and Reactor Centre,
East Kilbride, Glasgow G75 0QU, Scotland

Introduction

The Galloway plutons (Figure 1) in the south of Scotland belong to the Newer Granites of the Caledonian orogenic belt (Read, 1961). The plutons are of particular interest for two reasons, namely, that they show marked diversity in overall composition from basic diorites to muscovite granites, and also that the plutons exhibit striking petrological zoning. Our objective in this study was to understand the causes of the petrological variation both within and between the plutons, and hence to determine the origin(s) of the magmas and the processes by which they were subsequently modified. The Galloway region assumes some geological importance in that all plate tectonic models of the Scottish Caledonian place one or more subduction zones near the present locations of the plutons (see review by Holland *et al.,* in press).

Zoned plutons are common in cordilleran batholiths and the zoning may show a variety of geometric forms and compositional trends. The Galloway plutons show normal concentric zoning, that is, the variation trends from more basic margins to a core of more acidic rock. In some cases the variation is continuous whereas in other instances discontinuities in composition or field contacts indicate an intrusion comprised of multiple pulses, which themselves also show the normal compositional trend. In a review of the origins of normal zoning, Vance (1961) concluded that the primary cause was crystallization from the margin inwards, sealing in volatiles which could transfer alkalis and silica to the pluton interior. Our objective was to test this and other hypotheses regarding inter- and intrapluton compositional variation. We conclude that the evolution of the Galloway plutons is complex and includes migration of magma sources, magma mixing and fractional crystallization.

Variation in petrographic types

Only the three largest plutons in Galloway are considered in this study: these are the Loch Doon, Cairnsmore of Fleet, and Criffell-Dalbeattie intrusions (hereinafter referred to as Doon, Fleet and Criffell; Figure 1). They are approximately equal in size (20 ± 2 km long) but vary in range of composition. Detailed petrography and field relations are given by Gardiner and Reynolds (1932), Phillips (1956) and Parslow (1968); the following summary also includes some

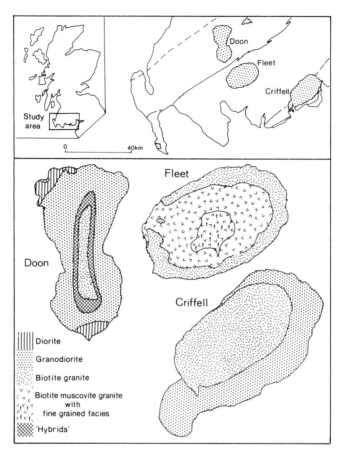

Figure 1 Location and petrology of the three main Galloway plutons. Note that the three petrological maps are not on exactly the same scales. (Map of Doon area from Gardiner and Reynolds, 1932; map of Criffell-Dalbeattie complex from Phillips, 1956 – both reproduced by courtesy of the Geological Society of London. Map of Fleet area from Parslow, 1968 – reproduced by courtesy of Scottish Academic Press)

more recent information.

Doon shows the sequence: diorite to granodiorite to biotite granite; Criffell varies from granodiorite to biotite granite with minor biotite-muscovite granite at the centre; Fleet has biotite granite passing into biotite-muscovite granite. The diorite of Doon is rather minor in area (12 km²) and is considered to be present as sheets in the roof of the main pluton. It contains orthopyroxene and often clinopyroxene, sometimes replaced by amphibole and, rarely, olivine may be found. The grain size is variable, some varieties showing chill textures and, as a whole, the 'diorite' is rather variable both petrographically and geochemically, including pyroxene-rich diorites, quartz diorites and quartz monzodiorites.

Granodiorite is the predominant rock in the Doon intrusion and also forms about half of the Criffell pluton (Figure 1). At Doon the granodiorite cross-cuts the diorites, and exhibits continuous normal zoning towards the central granite, as shown by a decrease in amphibole and plagioclase and an increase in quartz and alkali feldspar. At Criffell the granodiorite is markedly foliated in the outer parts and contains abundant xenoliths and patches of glomeroporphyritic amphibole. Protoclastic textures are common.

Biotite granites form the cores of both the Doon and Criffell intrusions. At Doon the boundary with the granodiorite appears transitional and Gardiner and Reynolds (1932) have mapped a zone of intervening hybrids. At Criffell the central granite forms a greater proportion of the pluton (Figure 1) and the boundary with the granodiorite is mapped on the basis of the incoming of alkali feldspar megacrysts (Phillips, 1956) since nowhere can a field contact be observed. The central granite is asymmetrically placed and lies in contact with country rock in the north-west of the pluton. Muscovite appears with biotite only in the most central zones of the Criffell intrusion.

Fleet comprises an outer coarse-grained biotite-granite which passes gradually into coarse-grained biotite-muscovite granite (Figure 1). The central part of the complex is formed by a fine-grained biotite-muscovite granite with prominent muscovite megracrysts. There is a mappable boundary between this and the coarse-grained variety (Figure 1).

The compositional range therefore decreases in the order Doon, Criffell, Fleet, and the granitoid types are progressively more evolved in the same order. It will be shown in the following sections that this order correlates well with absolute age and with isotopic composition.

Major oxide and normative mineral trends

THE THORNTON AND TUTTLE INDEX

This parameter in granites is the sum of normative quartz, orthoclase and albite, and is a good summary index of bulk composition. All three plutons exhibit good concentric trends in this index, but here only variations in the Criffell intrusion will be considered in detail. A contour map (Figure 2a), based on 180 data points, was prepared using an objective computer algorithm (a simple inverse distance weighting interpolation). The zoning pattern appears in general to be continuous, but in detail gradients are quite variable as is obvious from the contour spacings. This may be analysed by means of an approximation to the first spatial derivative, that is, a contour map of the *rate of change* of the index over the pluton (Figure 2b), as described by Stephens and Halliday (in preparation). A ridge in the first spatial derivative represents a particularly steep gradient, and a continuous ridge may be followed within the pluton, the trace of which approximates closely to the field boundary between granodiorite and granite (cf. Figure 1). The first derivative ridge demonstrates that the boundary between these two components is a discontinuity and that the pluton is composite. However, the fact that the magnitude of the gradient is markedly variable means that the compositional difference between the magma pulses is itself variable. The simplest explanation of this is that some interaction has taken place across the boundary between the pulses and we suggest

11

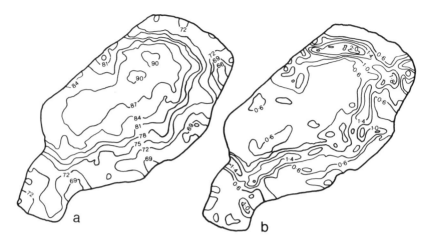

Figure 2 (a) Contour map of the Thornton and Tuttle Index over the Criffell pluton.
(b) Approximation to the first derivative of (a) with values of less than 0.6 excluded

that the amount of interaction correlates inversely with the magnitude of the first spatial derivative. This conclusion is supported by the initial strontium isotopic variations discussed later.

The Criffell pluton thus appears to be made up of an early granodioritic (low-index) pulse and a later granitic (high-index) pulse, and much *in situ* interaction may have taken place between the two. The central granite appears then to have undergone some fractional crystallization as shown by an increase in the index (Figure 2a). Doon shows a more continuous variation in the granodiorite-granite part of the pluton. In Fleet the range of variation is much less, although the absolute values of the index are higher than in either Doon or Criffell (54–86 in Doon, 66–90 in Criffell, and 86–94 in Fleet). In Fleet the lowest values tend to occur towards the outer margin.

NORMATIVE CORUNDUM-DIOPSIDE VERSUS SiO_2 TREND

A trend of diminishing diopside and increasing corundum with SiO_2 of whole rocks has been shown to hold for many calc-alkaline plutons, batholiths and extrusives from a wide range of tectonic settings (Cawthorn *et al.*, 1976). The same trend is shown by the samples from the Galloway plutons which show a continuous change in values from the Doon diorites to the Fleet muscovite-rich granites (Figure 3). Cawthorn *et al.* (1976) discussed the likely origin(s) of this fundamentally important trend and concluded that amphibole fractionation was the dominant process. With the aid of Sr-isotope data we show that this process alone cannot account for the diversity of composition in the Galloway plutons, and that mixing of magmas is the likely predominant process.

THE AFM TREND

The Galloway pluton data describe trends on this diagram typical of many other calc-alkaline intrusive provinces with similar petrographic range (Figure 3).

12

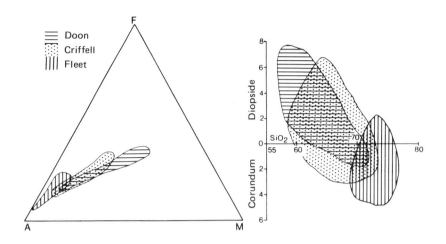

Figure 3 AFM and normative corundum or diopside vs. SiO$_2$ plots for the combined data for the three Galloway plutons

Strontium-isotopes

Isotopic studies have been made of all three plutons and their host metasediments. Reconnaissance U-Pb zircon and common Pb studies are reported in Pidgeon and Aftalion (1978), Blaxland *et al.* (1979) and Halliday *et al.* (1979).

Rb-Sr mineral-whole rock ages for the plutons are as follows: Doon, 408 ± 2 Ma, Criffell, 397 ± 2 Ma, Fleet, 392 ± 2 Ma. In Figure 4 (^{87}Sr/^{86}Sr)$_t$ values (where t is the respective age given above) are plotted against the present-day ^{87}Rb/^{86}Sr ratio for whole rock samples from the three plutons. Also plotted are the values of (^{87}Sr/^{86}Sr)$_t$ for 13 Southern Upland Lower Palaeozoic sediments covering the main spectrum of compositional types. The vertical line symbol connects the calculated values that a sediment sample had 408 and 392 Ma ago. The weighted average 397 Ma ago is marked as a star in the diagram. A similar arrangement of points is found on a plot of (^{87}Sr/^{86}Sr)$_t$ vs. 1/Sr (not shown), with the exception that the sediments display more scatter. It is clear that the overall initial Sr-isotopic composition was increasingly radiogenic in the order Doon to Criffell to Fleet, that is, with successive emplacement. From Figures 4 and 5 it is apparent that the initial ratios display an overall increase with increasing Rb/Sr. Hence, as the Rb/Sr ratios increase inwards within each pluton, so also do the initial ratios with minor exceptions (Figure 6).

At the present erosion level the older sediments in the immediate vicinity of each pluton contain relatively radiogenic Sr. It is clear, therefore, that wall-rock reaction has been minimal and is not the cause of the lower initial ratios at the margins. A suite of xenoliths, collected from Craignair Quarry (Figure 6) within 200 m of the contact between the central Criffell granite and metasediments,

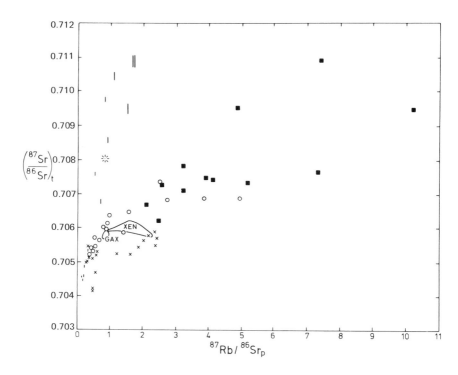

Figure 4 Plot of $(^{87}Sr/^{86}Sr)_t$ against present-day $^{87}Rb/^{86}Sr$. Symbols and values of t are: crosses = Doon, 408 Ma; circles = Criffell, 397 Ma; filled squares = Fleet, 392 Ma; vertical lines = sediments (see text)

were all nearly in isotopic equilibrium with the local magma (Figure 5). As these represent the range of xenolith types found at this locality, it is apparent that a greater degree of assimilation of xenoliths in the centre of the pluton is not the cause of the isotopic zoning. The most altered Criffell sample, 8478, plots (Figure 5) with an apparently low initial ratio. A metasomatite from the centre of the Criffell intrusion (sample 219) also plots off the main trend of data points. These results suggest that the effect of subsolidus modification is to produce a scatter in the isotope data. Hence, the systematic trend of the other points in Figure 5 is not attributable to subsolidus alteration (which, in any case, is often minor in the Southern Uplands granites).

The approximately linear correlation of initial ratio with $(^{87}Rb/^{86}Sr)_p$ in the granodiorites and outer granites of Criffell (Figure 5) corresponds to an apparent age of 512 ± 13 Ma before present, but is not attributable to time-dependent variations in a single source region since the magmas were emplaced in a sequence in the direction basic to acidic from the margins inwards. Melting of a common source would be expected to yield the reverse emplacement sequence. The same applies to the other plutons.

14

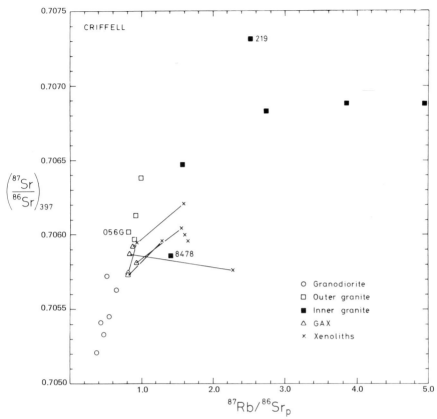

Figure 5 Plot of $^{87}Sr/^{86}Sr$ initial ratios against present-day $^{87}Rb/^{86}Sr$ for samples from the Criffell intrusion

Models that explain the isotopic variations simply in terms of fractional crystallization are of two types:

1. Fractionation *in situ* of solid phases which were isotopically disequilibrated from the melt phase by preferential contamination of the latter by radiogenic strontium.
2. Fractional crystallization over sufficient time (tens of millions of years) such that $^{87}Sr/^{86}Sr$ in the melt increased significantly with successive deposition of cumulate phases enhancing the residual Rb/Sr ratio.

Model 1 is difficult to apply, chiefly because of the large amounts of contamination required. The second model also has problems, insofar as the kinds of $^{87}Sr/^{86}Sr$ variation considered here require unrealistically high Rb/Sr ratios over long periods of time. For example. the Criffell magma would need to have initially crystallized out phases with $^{87}Sr/^{86}Sr < 0.7052$ (lowest Criffell) and finally crystallized with $^{87}Sr/^{86}Sr \simeq 0.7069$ (highest magmatic Criffell). Allowing a

15

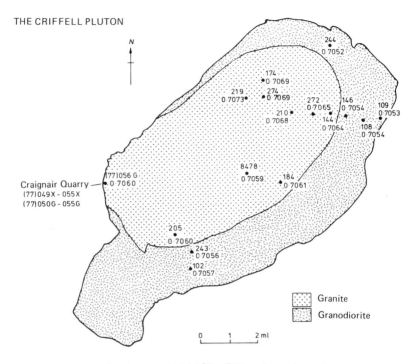

Figure 6 Map showing distribution of initial $^{87}Sr/^{86}Sr$ ratios in the Criffell intrusion

time-span of 20 Ma prior to emplacement, the samples with highest magmatic $(^{87}Sr/^{86}Sr)_{397}$ in Criffell would need to have evolved their Sr isotopic composition in an environment with an average $^{87}Rb/^{86}Sr$ ratio (present-day equivalent) of > 5.94. This value is higher than the highest $^{87}Rb/^{86}Sr$ value determined from Criffell and the melt cannot have had such an enhanced $^{87}Rb/^{86}Sr$ ratio until the final stages of solidification.

The isotopic variations are therefore taken to reflect mixing of magmas derived from differing source types. On the one hand this is probably a metasediment partial melt, and on the other a more primitive mafic source with low Rb/Sr and $^{87}Sr/^{86}Sr$. Each pulse of magma (granodiorite and granite in the case of Criffell) is itself a hybrid to a greater or lesser extent and the magma pulses became progressively enriched in the metasedimentary component, both in the overall interplutonic cycle and during the time-span of emplacement of each pluton. Mixing of magmas is also evident *in situ*. For example, the initial ratios of Criffell increase in the granodiorite from NE to SW, whereas in the granite they decrease from NE to SW, resulting in a steeper 'jump' in initial ratio in the NE than in the SW (Figure 6). The chemical variations display similar effects. A likely mechanism for this is that mixing across the intrusive boundaries was more thorough in the SW than in the NE. Intrusion of the granodiorite magma may have aided mobilization of melts from higher levels in the crust, which intruded the granodiorite magma at depth and rose into its hot, plastic core. There are sections of each

16

pluton which are chemically variable but are more or less isotopically uniform. This is attributed to fractional crystallization, possibly *in situ.*

Conclusions

Although only Criffell has been considered in detail, data from all three plutons enable us to arrive at the following conclusions regarding the Galloway plutons:

1. We consider the source of the *diorites and granodiorites* to be primarily a 'basaltic' part of the lower crust and/or upper mantle. However, the *granites* at the centre of Criffell and Doon and the whole of Fleet have a high proportion of metasedimentary melt component. The metasedimentary source may be lower crust but the data do not preclude Lower Palaeozoic sediments.
2. The granodiorite and granite magmas have mixed to some extent to form hybrid boundary zones in Criffell and Doon and give rise to continuous trends on variation diagrams.
3. Fractional crystallization processes may relate in part the granodiorites of Doon to the main diorites. Continuous zoning in the granodiorite and granite of Doon, the granite of Criffell and the granite of Fleet may be ascribed to fractional crystallization.
4. Contamination by assimilation of wall-rocks *in situ* has played an insignificant role in the large-scale modification of magmatic compositions in the interiors of the plutons.
5. Concentric zoning in the Galloway plutons is a complex function of migration of melt source, magma mixing and fractional crystallization.
6. Spatial zoning and temporal changes to more sedimentary-derived melts point to an ascending melt zone incorporating melts from higher levels in the crust with time.

ISOTOPE AND TRACE ELEMENT EVIDENCE FOR THE ORIGIN AND EVOLUTION OF CALEDONIAN GRANITES IN THE SCOTTISH HIGHLANDS

R.J. PANKHURST
British Antarctic Survey, c/o IGS (NERC),
64-78 Gray's Inn Road, London WC1X 8NG, UK

Introduction

Application of radiogenic isotope geochemistry to the petrogenesis of granitic magmas has been a major field of investigation since the foundations of Sr-isotope geology were laid down by Faure and Hurley (1963). In general, it has been shown that granites display a very wide range of initial Sr and Pb-isotope compositions, open to an equally wide range of interpretations. Thus, Faure and Powell (1972) summarized the Sr-isotope data by recognizing three groups with rather arbitrary boundaries: (a) granites with initial $^{87}Sr/^{86}Sr$ ratios falling within the range of mantle-derived basic magmas of the same age (i.e. < 0.706 at the present day), (b) granites with initial $^{87}Sr/^{86}Sr$ ratios equal to or higher than those predicted for old continental crust (i.e. > ca 0.718 at the present day), (c) granites with initial $^{87}Sr/^{86}Sr$ ratios between these two fields. It has not been possible to use this classification confidently as a simple rule whereby the origin of individual granites may be ascribed to: (a) melting in the mantle, (b) melting in the crust, and (c) hybridization or contamination, respectively, because the range of potential source regions defies such a simple categorization. Thus, although the parent igneous rocks of the broadly 'granitic' Archaean shield areas fall mostly into group (a), and thus cannot have been derived from already old crust with high Rb/Sr ratios (e.g. Moorbath, 1975), other geochemical considerations favour an origin by remelting of juvenile crustal materials, extracted from the mantle shortly beforehand (e.g. O'Nions and Pankhurst, 1978; Hamilton et al., 1979). At the other extreme, metasomatic processes and/or alteration may sometimes be responsible for granitic rocks acquiring crustal isotope characteristics which mask any information about their true magmatic source regions (e.g. Brooks and Compston, 1965; van Breemen et al., 1975).

Isotope geology has been of greatest significance in detailed regional studies of granites. Probably the best documented area is the Western Cordillera of the USA (Kistler and Peterman, 1973; Doe and Delevaux, 1973; Zartman, 1974; Amstrong et al., 1977). Here, there is good evidence that both the Sr- and Pb-isotope compositions of Mesozoic batholithic granites is controlled by the nature of underlying crustal rocks. 'Mantle type' initial $^{87}Sr/^{86}Sr$ ratios (< 0.706) and $^{206}Pb/^{204}Pb$ ratios (> 19) are characteristic of granites in the coastal belt where the continental crust consists of juvenile eugeosynclinal sediments and volcanics. A sharp transition to more-radiogenic Sr and 'old', less-radiogenic Pb occurs eastwards along a line thought to correspond to the edge of the underlying Precambrian continental shield.

Although such data clearly indicate the importance of crustal involvement in orogenic granite magmatism, there is no general agreement over the extent to which they reflect either crustal source regions or merely crustal contamination of mantle-derived magmas. The greatest advances in this argument in recent years have been: (a) the recognition of related chemical and isotopic variations which enable discrimination between 'S'-type granites, probably derived from sedimentary crustal sources, and 'I'-type granites with 'primitive' igneous origins (Chappell and White, 1974; see also Beckinsale, this volume, and Coleman, this volume), and (b) the recognition that 'I'-type granites produced at destructive plate margins are chemically and isotopically very closely related to andesite volcanism and may share a common mantle origin (see Thorpe and Francis, this volume). However, there are orogenic environments not easily related to these models where the 'crustal involvement' dilemma remains to be fully resolved, and the British Caledonides represent one of the best known examples of this.

The Caledonian granites of the Scottish Highlands

The metamorphic belt of the Caledonides was formed during late Precambrian and Lower Palaeozoic times along the south-eastern margin of the North American—Scandinavian shield, bounded by a proto-Atlantic or Iapetus Ocean (Dewey, 1974; Phillips et al., 1976; Wright, 1976). Magmatism and tectonic activity are thought to have been largely controlled by subduction of ocean floor north-westwards and reached a climax in Ordovician-Silurian times as Iapetus disappeared and progressive collision with a SE Foreland continent occurred.

Granites were emplaced into the metamorphic Caledonides over a considerable period of time and under a variety of regional conditions. The classification proposed by Read (1961) is still adhered to today, although geochronological and geochemical studies, now long overdue, will probably lead to considerable refinement. The Older Granites are those affected by Caledonian deformation and metamorphism, now converted to augen gneiss. The largest Older Granite bodies are the Carn Chuinneag-Inchbae intrusion of Ross-shire and the Ben Vuirich granite of Perthshire. However, these are volumetrically insignificant compared to the undeformed (post-tectonic) Newer Granites. These were divided by Read into two groups: the Forceful Newer Granites, emplaced by active injection into the country rocks with formation of contact migmatites and veining (e.g. Rogart, Cluanie, Foyers), and the Last 'permitted' Granites, 'passively' emplaced in ring complexes associated with tensional faulting and cauldron subsidence (e.g. Ben Nevis, Etive, Lochnagar).

AGE

Radiometric dating has been applied successfully to the Older Granites, using Rb-Sr whole-rock and U-Pb zircon methods. Ages of 555 ± 10 Ma for Carn Chuinneag (Pidgeon and Johnson, 1974) and 514 ± 7 Ma for Ben Vuirich

(Pankhurst and Pidgeon, 1976) seem consistent with their structural history and an early Ordovician age (*ca* 480 Ma) for the main regional metamorphism. In the Buchan region of North-east Scotland there is a distinctive group of post-metamorphic biotite-granites (Strichen, Longmanhill, Aberchirder and other small outcrops associated with the Newer Gabbros) which, in terms of petrography and geochemistry, seem to be comparable with the Older Granites. These have been dated at *ca* 460 Ma (Pankhurst, 1974; Pidgeon and Aftalion, 1978). On the other hand, the 456 ± 5 Ma Glen Dessary syenite of western Inverness-shire (van Breemem *et al.*, 1979) is syn-tectonic and chemically more like the succeeding Forceful Newer Granites.

Read (1961) argued that most of the Forceful Newer Granites were emplaced, uplifted and eroded prior to deposition of the Lower Old Red Sandstone (LORS) strata involved in the Last Granite ring complexes. As affirmed by Watson (1964) and Watson and Plant (1979), the differences of intrusive style among the Newer Granites must represent greater depth and country rock temperatures for forceful emplacement. This means that radiometric determination of the age difference is difficult to establish, since it is to be expected that K-Ar mica ages for the forceful granites would reflect post-intrusive cooling patterns in the country rocks (Dewey and Pankhurst, 1970). Thus, K-Ar ages for most forceful granites are indistinguishable from those of the permitted granites (390–420 Ma, Brown *et al.*, 1968). Recent dating using the U-Pb zircon method (Pidgeon and Aftalion, 1978) has confirmed this range in a general way, although the fact that these ages have been calculated from discordant zircon arrays often leads to a lack of precision and, it has to be re-emphasized, a strong possibility that even slight subsequent isotopic disturbance will yield too young an inferred age (e.g. 365 Ma for the Ratagain granite). The chief exceptions are the Strontian complex, where zircons from a xenolith and enclosing tonalite gave an age of 435 ± 10 Ma, i.e. late Ordovician, and the Borolan alkaline complex now dated at 429 ± 3Ma (van Breemen *et al.*, in press). It is quite probable that most of the forceful granites are of similar age to these. Moreover, it seems that even the LORS granites are ≥400 Ma, significantly older than the now well-dated Shap granite (393 ± 5 Ma, Pidgeon and Aftalion, 1978; Wadge *et al.*, 1978) and thus probably pre-Devonian.

CHEMISTRY

Since the overall reconnaissance work of Nockolds and Mitchell (1948) and Mercy (1963), there have been chemical studies of the Strontian (Sabine, 1963), Foyers (Marston, 1971) and Ben Nevis (Haslam, 1968) complexes. These have mostly concentrated on major element chemistry and have established that the Newer Granites are high Ca and Sr calc-alkaline rocks with continuous chemical variation between the various rock-types, usually ascribed to crystal fractionation. Modern trace element chemistry has been conspicuously lacking, but recently Watson and Plant (1979) and Simpson *et al.* (1979) have published U analyses. Brown (1979b) and Brown and Locke (in press) have given some new geochemical and geophysical data. In all these recent studies, it has been proposed that the Older Granites and Forceful Newer Granites are related in 'merging' geochemically and geophysically with their background country rocks, whereas the Last Granites show contrasts with respect to U, K/Rb, initial $^{87}Sr/^{86}Sr$ and gravity anomalies. It is inferred that the former groups are largely crustal melts whereas the latter are mantle melts contaminated as they rise through the lithosphere by 'scavenging' K, Rb, U and ^{87}Sr from the wall rocks.

INHERITED ZIRCON

One of the most remarkable features of these Caledonian granites is that many contain at least two populations of zircon crystals. In addition to euhedral zircons, there are often darker, euhedral or broken and rounded crystals, sometimes forming cores to later igneous growth, which have U-Pb systems established 1000–2000 Ma ago (Pidgeon and Johnson, 1974; Pankhurst and Pidgeon, 1976; Pidgeon and Aftalion, 1978). These can only be xenocrysts incorporated into the Caledonian granitic magmas from older crustal source rocks. They seem to occur almost exclusively in granites north-west of the Highland Boundary Fault, so that Pidgeon and Aftalion (1978) concluded that these were derived by partial melting of Proterozoic (reworked Lewisian) crust at depth, whereas Caledonian granites in the Southern Uplands and the Lake District were derived by partial melting of Palaeozoic source rocks. Halliday et al. (1979) have challenged the uniqueness of this conclusion, pointing out that the zircon xenocrysts could equally have come from secondary sources (i.e. sediments derived from Proterozoic crust) and that a hybrid origin or crustal assimilation were strong alternative possibilities to crustal melting.

STRONTIUM-ISOTOPE GEOLOGY

Whole-rock initial $^{87}Sr/^{86}Sr$ ratios are summarized in histogram-form in Figure 1. Only recent high-precision measurements are included, mostly not yet published. The chief features are immediately obvious.

The Older Granites and post-tectonic granites of North-east Scotland have a wide range of initial $^{87}Sr/^{86}Sr$ ratios, with most greater than 0.710. With the exception of the Glen Dessary syenite previously noted, all those analysed have shown a high proportion of inherited zircon. These features, together with high Rb/Sr ratios (typically $\geqslant 1$) and the biotite-K feldspar-rich nature of these rocks, has lead to a consensus view that the magmas were indeed produced by crustal anatexis (Bell, 1968; Pankhurst, 1974; Pankhurst and Pidgeon, 1976). The Forceful Newer Granites also display a significant range, but with much lower initial ratios (0.704–0.709) which barely overlap with the pre-450 Ma granites. They have a pronounced maximum at the lower end of this range, indistinguishable from the 0.704–0.706 peak from the Last Granites (chiefly Etive). These low values are typical of 'I-type' granites generated at destructive plate margins, and would normally be considered to indicate a subcrustal source region. Thus, despite the geophysical grouping of the Forceful Newer Granites with the Older Granites, and the fact that most of these contain inherited zircons whereas none of the Last Granites do, the Sr-isotope evidence would seem to require a contrasting origin for the Newer Granites as a whole compared to the pre-450 Ma granites.

It is important to realize that low initial $^{87}Sr/^{86}Sr$ ratios do not necessarily

21

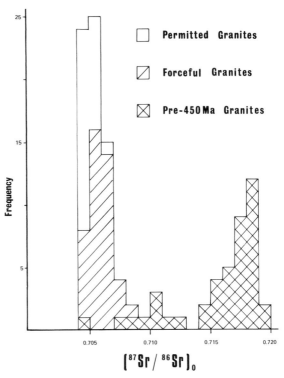

Figure 1 Histogram of initial $^{87}Sr/^{86}Sr$ ratios of Scottish Caledonian granites. Pre-450 Ma granites: Carn Chuinneag (Pidgeon and Johnson, 1974); Ben Vuirich and Dunfallandy Hill (Pankhurst and Pidgeon, 1976); Strichen, Longmanhill and Aberchirder (Pankhurst, 1974); Glen Dessary (van Breemen *et al.,* 1979). Forceful Newer Granites: Strontian, Foyers, Cluanie, Bonar Bridge, Helmsdale, Rogart, Ballachulish, Strath Ossian, Moor of Rannoch and Ross of Mull (Halliday *et al.,* 1979; Pankhurst, unpublished). Last Granites: Cruachan, Starav, Ben Nevis, Lochnagar, Hill of Fare (Halliday *et al.,* 1979; Brown, 1975; Pankhurst, unpublished)

imply direct derivation from the mantle. Other suitable sources would be young igneous rocks (within the continental crust, underplating it, or in subducted lithosphere), immature sediments derived therefrom or even old crustal rocks with very low Rb/Sr ratios. In this context it seems most desirable to investigate the cause of the significant variation of the Forceful Newer Granites, which might be due to contamination of a primitive 'I-type' magma with upper crustal materials, or a range of compositions existing within the source at the time of melting. The remainder of this paper is concerned with a detailed study of two of these forceful complexes – Strontian and Foyers.

The Strontian and Foyers complexes

These two intrusions are well known – Strontian from its high content of Sr, resulting in crystallization of the minerals strontianite and celestite in late veins, and both from the suggestion of Kennedy (1946) that they once formed a single pluton which was disrupted by a 100 km sinistral displacement on the Great Glen Fault. Each has a crude annular outcrop pattern, truncated by the fault, with an outer tonalite and an inner granodiorite, both cut by a late biotite-granite or adamellite (Figure 2). This sequence is, in fact, typical of many Newer Granites.

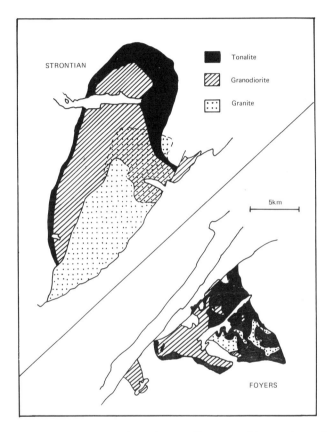

Figure 2 Geological sketch map of Strontian and Foyers granitic complexes after Marston (1971). *NB*. The two are shown side-by-side for convenience. Data discussed here disprove the hypothesis that they were originally parts of a single intrusion

The permitted granites 'passively' emplaced generally have a more perfect annular structure with apparently gradational internal contacts, although even here successive intrusion may be the rule rather than *in situ* differentiation (see Holder, this volume).

The only rock unit for which there is a satisfactory radiometric date is the tonalite at Strontian — U-Pb analyses of zircons extracted from a xenolith have given an age of 435 ± 10 Ma, in close agreement with similar results from the

tonalite itself (Pidgeon and Aftalion, 1978). Zircon from the Foyers granodiorite*
and the Strontian biotite-granite (Halliday *et al.,* 1979) give much younger apparent ages of around 385 Ma, but these are younger than existing K-Ar biotite ages
averaging about 400 Ma (see Brown *et al.,* 1968). Whereas the Strontian tonalite
data are essentially concordant, the remaining zircon ages were derived by extra-
polation of reverse discordia caused by the occurrence of inherited Proterozoic
zircons in the population, and it seems that these have been significantly rotated
by post-crystallization disturbance (see Pidgeon and Aftalion, 1978).

Major element data have been published for both intrusions and are summarized
by Marston (1971). There is a close correspondence between rock compositions
at Strontian and their equivalents at Foyers, although the latter shows a more
extended range. A continuous increase in K and Si from tonalite to adamellite
is accompanied by a simultaneous fall in Ca, Mg, Fe, Ti and Al. This, together
with the observed changes in modal mineral contents and the observation that
normative compositions in the quartz-plagioclase-alkali feldspar system trend
towards the 2 kb ternary eutectic, was taken to indicate a liquid line of descent,
hornblende, plagioclase and ore minerals being removed by fractional crystallization. Such differentiation must have occurred below the present level, each
successive phase being intruded while the previous one was still hot, or even still
partly liquid. Final emplacement was also accompanied by some contamination
at the margins of the intrusion and a late-stage alkali metasomatism (Marston,
1971).

A common problem in geochemistry is that major element relationships, such
as those described above, do not unambiguously distinguish between probable
petrogenetic mechanisms. Similar trends can be expected from crystal fractionation, variable degrees of partial melting of a homogeneous source, partial melting
of an inhomogeneous source in which variations were originally due to igneous
differentiation and even, in some circumstances, contamination of a basic magma
by assimilation of crustal rocks. In order to constrain these possibilities further,
fresh samples from the Strontian and Foyers complexes have been analysed for
major elements, selected trace elements and radiogenic isotope contents.

ISOTOPE SYSTEMATICS

Rb-Sr data are plotted in the conventional isochron diagram in Figure 3. Three
groupings are apparent. The adamellite unit at Foyers yields data points which
are aligned along a 415 Ma reference isochron, which is taken to indicate a uniform
initial $^{87}Sr/^{86}Sr$ ratio of *ca* 0.7045 at the time of crystallization. The remaining
rocks, however, show marked variation in this parameter. The tonalites and
granodiorites at Foyers have initial $^{87}Sr/^{86}Sr$ ratios (assuming an age of 430 Ma)
which range from 0.7060 to 0.7081, and those at Strontian range from 0.7053 to
0.7072 (the latter value occurring in the adamellite which is thus clearly distinct
from its counterpart at Foyers). Moreover, there are crude correlations with

---

*The sample from Foyers analysed by Pidgeon and Aftalion (1978) was described as a grano-
diorite. However, the locality given is within the tonalite outcrop as mapped by Marston
(1971). Moreover, according to Halliday *et al.* (1979) the Sr content is 1240 ppm, so that it
would plot with the other tonalites in Figures 5 and 6 of the present study.

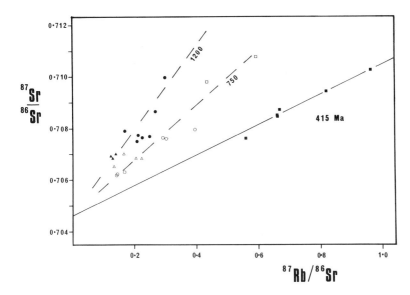

Figure 3 Rb-Sr systematics for Strontian and Foyers (including four data points from Halliday *et al.*, 1979). Open symbols = Strontian; closed symbols = Foyers; triangles = tonalite; circles = granodiorite; squares = adamellite. Only the Foyers adamellite points conform to a late-Caledonian isochron age

Rb/Sr ratios, resulting in 'pseudoisochrons' of *ca* 1200 Ma for the early units at Foyers and *ca* 750 Ma for all rocks at Strontian (both estimates subject to considerable uncertainty). These pseudoisochrons may be interpreted in either of two general ways (Pankhurst, 1977): (a) inheritance of correlated Sr-isotope and Rb/Sr variation from an inhomogeneous source region, or (b) mixing of two or more components, one with low $^{87}Sr/^{86}Sr$ and low Rb/Sr and another with high values for both parameters.

In either case, the ultimate age of the crustal component (be it source or contaminant) *may* be approximated by the pseudoisochrons. It is significant that the results for Foyers tonalites and granodiorites are comparable with the 1000–2000 Ma ages of inherited zircons in this and in other Newer Granites, suggesting that the same crustal unit is responsible. The younger pseudoisochron age for Strontian may represent either a younger crustal component (? Morarian rather than Grenvillian) or else a uniformly higher fractionation of Rb relative to Sr during magma genesis (e.g. due to smaller degrees of melting or more extensive crystal fractionation). The latter possibilities seem unlikely in view of other geochemical comparisons between the two complexes.

A crucial and unexpected result concerns the most evolved rocks at Foyers: the adamellites. On the crustal melting model (a), these would have to be derived from the least evolved source compositions, i.e. those with the lowest Rb/Sr ratios, but involving a very great increase in Rb/Sr. This requires a fundamentally different melting process, or else a completely different source region, from the other rocks. Alternatively (b), they must have formed with no significant crustal contamination compared to the earlier members of the complex. The simplest test of the last hypothesis would be an analysis of the zircons in the Foyers adamellite in order to detect any inherited component. In the absence of such information we must rely upon evidence from other isotope systems and trace elements concerning magma sources and processes.

Pb contents and Pb-isotope compositions were determined for representative samples from the Foyers complex. Pb contents increase from 12 ppm in the tonalite to 23 ppm in the adamellite, which compares with the amount of increase in Rb. Using U contents determined for the same samples (R.S. Gollop, personal communication), the composition data were corrected for radiogenic decay since crystallization. Initial $^{206}Pb/^{204}Pb$ ratios (17.0–17.3) and $^{207}Pb/^{204}Pb$ ratios (15.3–15.4) are plotted in Figure 4 together with recent Lewisian Pb-isotope

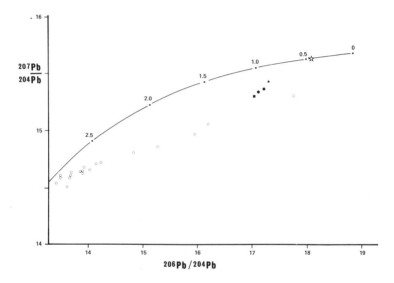

Figure 4 Initial Pb-isotope systematics for Foyers (symbols as in Figure 3). The curve is Cumming and Richards' (1975) model for the mantle source of ore leads with a star at 430 Ma. The open circles are present-day Pb-compositions of Lewisian gneisses (Chapman and Moorbath, 1977). Correction of the latter to 430 Ma would move points slightly to the left, but this cannot be done since U and Pb contents are unavailable

data (Chapman and Moorbath, 1977) and Cumming and Richards' (1975) linear-increase model for the mantle source of ore leads. It is clear that the results from Foyers form a fairly tight cluster, well removed from both the composition of 430 Ma mantle Pb and the predominant field of Lewisian gneisses. Single-stage model ages are 720–770 Ma. This would require a primordial source region with a significantly lower μ-value ($^{238}U/^{204}Pb$) than normal mantle, but it in any case seems unlikely that a single-stage model is appropriate. A more reasonable interpretation would involve mixing of Lewisian-type crustal Pb with juvenile Caledonian Pb. It is not possible to distinguish whether this mixing occurred in

26

the source region prior to magma generation (e.g. by sedimentation) or subsequently due to contamination. Despite the small spread of Pb isotope compositions, that in the adamellites is actually closer to the crustal component than is that in the tonalite — the reverse of the case for Sr-isotope compositions. This suggests independent behaviour of the two isotope systems and, probably, a multi-component mixing process.

Preliminary Nd-isotope data, in collaboration with P.J. Hamilton and R.K. O'Nions, shows the same general features as Pb, i.e. initial $^{143}Nd/^{144}Nd$ ratios of 0.5118 and model ages based on a chondritic source of 500–1000 Ma.

Further constraints on the nature of end-members in possible mixing models may be obtained from other geochemical considerations.

TRACE ELEMENT GEOCHEMISTRY

The variation of some selected elements in these two intrusions is shown in Figure 5. As demonstrated by Marston (1971), K_2O increases and CaO decreases uniformly for both, with increasing SiO_2. This applies also to Sr and Ce, which decrease by factors of 5 and 3, respectively, over the range considered here. Rb increases at the same rate as K, i.e. slowly compared to the fall in Ca and Sr. However, selected inter-element relationships shown in Figure 6 reveal some persistent differences between the two intrusions. Thus, K/Rb is lower at Strontian (*ca* 370) than at Foyers (*ca* 420) and Ca/Sr is slightly lower. Chondrite-normalized

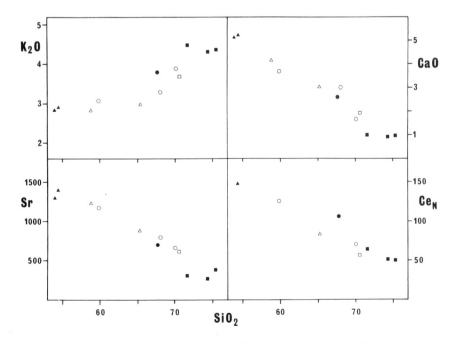

Figure 5 Variation of selected elements versus SiO_2 for the Strontian and Foyers complexes. Note the compatibility of Sr and Ce (symbols as in Figure 3)

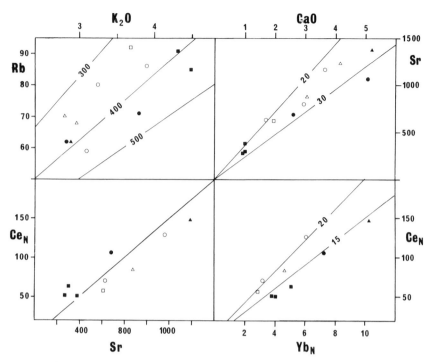

Figure 6 Selected inter-element relationships for the Strontian and Foyers complexes (symbols as Figure 3)

Ce/Yb ratios are higher at Strontian throughout the range tonalite to granite. These differences alone are enough to establish that, although magmas evolved along parallel paths in the two intrusions, each has its own trace element characteristics, and they cannot be separated portions of a single pluton.

Several notable features emerge from the full chondrite-normalized rare earth element (REE) patterns (Figures 7 and 8). They confirm the strong light-to-heavy REE fractionation suggested by the Ce/Yb ratios, which are 2–3 times higher than in typical calc-alkaline granites from destructive plate margins (e.g. Thorpe and Francis, this volume). In the case of Strontian, patterns are essentially subparallel for all rock types but without the occurrence of Eu anomalies which would be expected from feldspar fractionation. Such anomalies are present in the Foyers rocks, slightly positive in the tonalite and granodiorite (Eu/Eu* ~ 1.1) but negative in the adamellites (Eu/Eu* ~ 0.7). The patterns for Foyers are not subparallel, despite the constancy of Ce/Yb ratios, but show progressively greater depletion of Dy and Er. Nevertheless, the smooth variations leave no doubt that all rock types are genetically related within each of the complexes (i.e. the Foyers adamellite is not a much later magma unrelated to the earlier units) although, once again, the individuality of the two intrusions is clearly established.

Most remarkable of all is that in both complexes total REE contents fall continuously from tonalite, through granodiorite, to granite — the reverse of

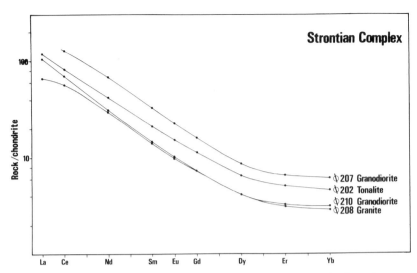

Figure 7 Chondrite-normalized rare earth element abundances for the Strontian complex (see Table 2 for analytical data)

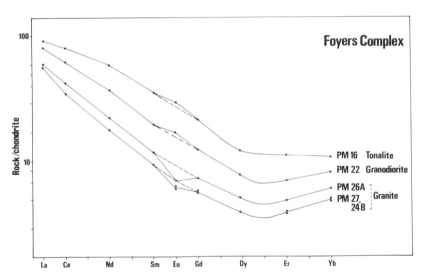

Figure 8 Chondrite-normalized rare earth element abundances for the Foyers complex (see Table 2 for analytical data)

normal igneous trends. This shows that the REE are behaving as compatible elements with bulk solid—liquid partition coefficients (D) greater than unity. In order to model trace element variations during partial melting or crystal fractionation it is necessary to know the values of D and the melt proportion (F) at each stage. The principles are beyond the scope of this paper and for these the reader is referred to Hanson (1978), who also summarizes partition coefficient data for individual mineral phases in granitic liquids. In the present case, the normal modelling procedure has been inverted, and the implied D values have been calculated from the data by estimating F from a four-fold increase in Th contents in the Foyers samples and a three-fold increase in the Strontian samples (R.S. Gollop, personal communication). Thus, it is assumed that the Foyers adamellites are produced by one-quarter the degree of melting which formed the tonalites (Model A), or by four times as much crystal fractionation (Model B). Results are given in Table 1, together with some calculated D values for possible crystallizing assemblages. The greater efficiency of crystal fractionation at depleting compatible elements is shown by the high D values implied by Model A at Foyers, which seem to exceed probable cumulus combinations. A further major problem with this model is that it is necessary to assume that the tonalites are virtually 100% melts in order to achieve finite values of D at all. At Strontian, at least 65% melting must be assumed for the most basic rock. Although it is possible that the tonalites are cumulates rather than true liquid compositions, this can only be partly true (because of the very small positive Eu anomalies) and this does not greatly affect the argument. Such an approach to total melting seems improbable, especially within the crust (necessary to account for the tonalitic chemistry of the melt). On the other hand, the difficulties in modelling the data by removal of observed phenocryst phases from a parent tonalite/granodiorite liquid are not too severe considering the uncertainties in quantitative estimates. It is concluded that such crystal fractionation is indeed the prime cause of chemical and petrological differentiation within these complexes. The characte-

Table 1 **Calculated distribution coefficients (D) for appropriate magma models**

Model details	Ce	Sm	Eu	Gd	Yb	Sr
A. Batch melting						
Strontian (F = 0.33)	2.8	3.0	2.7	2.7	2.7	2.4
Foyers (F = 0.25)	3.5	7.1	10.2	7.4	3.4	6.4
B. Crystal fractionation						
Strontian (F = 0.33)	1.7	1.8	1.7	1.7	1.7	1.6
Foyers (F = 0.25)	1.8	2.2	2.5	2.3	1.8	2.2
C. Possible mineral assemblages						
50% hornblende, 50% plag	0.9	4.0	3.6	5.0	4.2	2.2
25% hornblende, 72% plag	0.6	2.0	2.9	2.6	2.1	3.3
X	1.1	2.2	2.6	2.2	1.4	>3.1

Notes: A. F values apply to adamellite liquid, F = 1 for tonalite.
B. Taking tonalite as the primary magma.
C. X is the average modal composition of the Foyers tonalite estimated from Marston (1971, Figure 2: 65% plag, 5% K-feldspar, 3% quartz, 15% biotite, 10% hornblende, 2% apatite.

ristic differences between the two intrusions may reflect differences in the parent magmas or in crystallization paths (e.g. accumulation of a greater proportion of hornblende or minor phases at Foyers).

Analytical data for rare earth element contents of Strontian and Foyers complexes are given in Table 2.

Table 2 **Rare earth element contents for Strontian and Foyers complexes (ppm)**

Sample	Grid reference	La	Ce	Nd	Sm	En	Gd	Dy	Er	Yb
Strontian										
207	NM 794555	47.8	110	43.4	6.67	1.75	4.48	2.95	1.47	1.34
202	NM 779610	38.9	70.9	26.5	4.28	1.16	3.13	2.23	1.14	1.02
210	NM 858526	34.6	61.0	19.7	2.97	0.82	2.02	1.39	0.72	0.69
208	NM 812550	22.1	49.3	18.8	2.85	0.81	2.07	1.41	0.70	0.63
Foyers										
PM16	NH 545227	58.4	128	61.2	10.2	3.11	7.47	4.33	2.51	2.34
PM22	NH 498189	49.4	91.5	33.9	4.75	1.48	3.59	2.47	1.38	1.63
PM26A	NH 573183	33.5	55.1	17.4	2.47	0.48	1.79	1.41	0.86	1.10
PM27	NH 561183	31.1	44.5	13.8	1.85	0.42	1.34	1.01	0.63	0.87
PM24B	NH 571185	–	44.8	13.1	1.82	0.39	1.30	1.00	0.66	0.82

Determined by mass-spectrometric isotope dilution analysis.

PETROGENESIS

The arguments presented above for the Strontian and Foyers complexes lead to the following general conclusions:

1. Trace element (especially REE) patterns constitute independent evidence for the prime cause of petrological variation in these complexes. Differential partial melting of a homogeneous source is effectively ruled out, and strong support provided for the hypothesis of crystal fractionation of plagioclase, hornblende and minor phases from a parental magma. Fractionation occurred below the level of emplacement, which was accomplished by successive intrusion.

2. Over a comparable range of major element chemistry, each of these two complexes exhibits its own distinctive trace element characteristics, showing that the parent magmas and their evolutionary paths were also distinct. The two outcrops do not represent separated portions of the same pluton.

3. Variation in initial $^{87}Sr/^{86}Sr$ ratios (0.704–0.709) conflicts with derivation from a single batch of parental magma at each complex. The simplest explanation of this would be contamination of early liquids with radiogenic Sr (and Rb) from old crustal rocks. It must have been possible for late liquids to fractionate to adamellite, either with (Strontian) or without (Foyers) suffering contamination. The absence of obvious contamination effects in the trace element data suggests either selective addition of fugitive components

or a close similarity in most trace element features between the magmas and the crustal contaminant.

4. The composition of the parental liquids would have been close to that of tonalite or basic granodiorite (55–60% SiO_2, 2.5–3.0% K_2O, *ca* 1000 ppm Sr, 100 ppm Ce, and with $Ce_N/Yb_N \sim 15$–20). Such a liquid could not be derived as a voluminous calc-alkaline melt from mantle peridotite. If, as seems most probable, it existed as primary liquid, it would have to have been derived from a composition at least as evolved as basalt. Approximately 10% melting of typical tholeiitic eclogite would be a reasonable single-stage mechanism for producing such a magma. Direct partial melting of Lewisian-type crust is ruled out by the Pb- and Nd-isotope data, and of Moine or Dalradian sediments by the Sr-isotope data (Halliday *et al.*, 1979). Immature crustal sediments (greywacke), partially derived from the Lewisian, might also be a suitable source (Halliday *et al.*, 1979; Stephens and Halliday, this volume) but the existence of such rocks at depth north of the Highland Boundary Fault is conjectural.

Significance for British Caledonian granites

In terms of the nomenclature of Chappell and White (1974), there is a marked distinction between the pre-450 Ma granites and the Newer Granites of the British Caledonides. The former group exhibits most of the diagnostic features of 'S-type' granites: small intrusions, restricted range of truly granitic composition, predominance of biotite over hornblende, a tendency to two-mica granites and high initial $^{87}Sr/^{86}Sr$ ratios. These certainly appear to have been generated by melting of old crustal rocks. On the other hand, the more abundant, post-metamorphic Newer Granites are predominantly 'I-type': large, complex intrusions, diorite-to-granite, essential hornblende, low initial $^{87}Sr/^{86}Sr$ ratios, high Ca and Na/K. Both forceful and permitted Newer Granites are of this type with only *slight* 'S-type' tendencies for the former, suggesting a common origin by melting of juvenile basic crust at high pressures, or possibly of immature sediments. Until these two possibilities are resolved, the true extent of upper crustal contamination indicated by Sr-isotope data and, in particular, the inherited zircon component cannot be fully constrained, but it certainly seems to have had little effect on petrological variation.

Recent ideas on the origin of granitic magmas in the British Caledonides depend crucially on the interpretation of trace element and isotope data. Most workers seem agreed on a crustal source for the Older Granites and Ordovician granites of North-east Scotland, but opinions differ over the Newer Granites. Some favour anatexis for these also, although within a regionally variable or stratified crust (Leake, 1978; Pidgeon and Aftalion, 1978) whereas others prefer a subcrustal source for the Last Granites (O'Connor and Bruck, 1976; Brown, 1979; Simpson *et al.*, 1979). The latter hypothesis is based on a tendency to low initial $^{87}Sr/^{86}Sr$ ratios in the late granites (especially those of Ireland and England), coupled with geochemical features, such as low K/Rb, low Sr and high Rb and U contents, which are considered incompatible with partial melting of depleted basement such as the Lewisian. These features, which except for the Sr-isotope data would normally be associated with upper crustal rocks, have been ascribed to residual phlogopite in the subcrustal lithosphere and selective 'scavenging' during ascent of the magmas. Much weight has also been given to contrasts between the Last Granites, which exhibit gravity and magnetic anomalies (Brown and Locke, in press), and the Forceful Newer Granites which generally lack such expression and 'both geochemically and geophysically merge with their host rocks'. Thus, a crustal origin has been inferred for the Forceful Newer Granites, as for the Older Granites.

However, it is the Newer Granites of Scotland which represent the acme of Caledonian magmatism and until now there have been few reliable Sr-isotope data for these. The compilations of Halliday *et al.* (1979) and the present paper require significant revision to the above arguments. Firstly, it is clear that there is no major isotopic distinction between forcefully and permissively emplaced granites — both have low initial $^{87}Sr/^{86}Sr$ ratios compared with the Older Granites. Secondly, in ascribing distinctive trace element abundances directly to source composition, too little account has been taken of enrichment by magmatic processes. The present work has reaffirmed the importance of fractional crystal-lization of primary diorite or granodiorite liquids in causing the observed petro-logical and chemical variation (cf. Nockolds, 1941). Extensive separation of hornblende could even account for decreasing K/Rb ratios. The writer believes that the origins of both types of Newer Granite were essentially similar (by melting of juvenile crustal rocks rather than mantle) and that the only important differences between them are those related to depth and style of emplacement as proposed by Read (1961). Even the geophysical aspects may be explained in this way, for example by separation of the Forceful Newer Granite magmas from early, basic, cumulates as they were squeezed into place.

In conclusion, it is indeed refreshing to find that the pioneering ideas of Nockolds and Read are still so much alive in the light of modern geochemical research.

Note added in proof:
Pb isotope analyses for feldspar separates from 11 Scottish Newer Granites are presented by Blaxland *et al.* (1979). These constitute a more representative basis for discussion than the four whole-rock Pb analyses for Foyers given here. The results and their interpretation are in close agreement with the present paper: overall, they distinguish a slightly greater range of initial Pb-isotope compositions but no systematic differences between Forceful and Last Granites.

ACKNOWLEDGMENTS
I would like to thank G.C. Brown, J.F. Brown, R.S. Gollop, A.E. Halliday and Jane Plant for making unpublished data or manuscripts available to me, and S. Moorbath and S.W. Richardson for help in obtaining samples.

GRANITE MAGMATISM IN THE TIN BELT
OF SOUTH-EAST ASIA

R.D. BECKINSALE
Geochemical Division, Institute of Geological Sciences,
64–78, Gray's Inn Road, London WC1X 8NG, UK

The relationship between tin ore deposits and granite magmatism

The tin belt of SE Asia, which is about 3500 km long and 400–800 km wide (Figure 1), is probably one of the best examples of a *metallogenic province*. The world's annual tin production (about 230 000 tonnes) is currently valued at about 3×10^9 US dollars and about 60% of this production comes from SE Asia. More dramatically, it is estimated that this metallogenic province has yielded about three-quarters of the world's total tin production during the present century. Most of the tin ore in SE Asia is mined from Quaternary placers, both onshore and offshore, and a steadily rising proportion of the annual production is from offshore dredging operations in increasingly deep water. In 1978, for example, about 35% of Indonesia's tin production was derived from offshore operations. It is anticipated that with the introduction of giant dredgers, designed to work in waters more than 30 m deep, offshore operations will account for about half of Indonesia's tin production by the early 1980s. The tin ore mineral, cassiterite (SnO_2), has been *concentrated* in the alluvial, eluvial or residual gravels of these placer deposits by a fortunate combination of two factors: (a) deep tropical weathering due to the climate (cassiterite is dense and resistant to weathering); and (b) Quaternary topography in relation to changes in erosional base level (producing thick alluvial 'flats'). Current tin production from deep mining of primary tin mineralization is less significant, although a large number of small-scale operations do exist. The only underground mines with substantial current production are the Pahang Consolidated Mine at Sungei Lembing near Kuantan, Malaysia and the Kelampa Kampit Mine in Billiton, Indonesia. The primary tin mineralization is *spatially* associated with particular granites, for example: (a) the fairly common occurrence of hydrothermal vein swarms (quartz–cassiterite–wolframite) cutting both the granite and the country rock in cusps in the roof of granite intrusions, e.g. Mawchi Mine, Burma (Hobson, 1940); Hermingyi Mine, Burma (Clegg, 1944; Khin Zaw, 1978). (b) The occurrence of so-called 'disseminated deposits', in which cassiterite is disseminated more or less throughout extensively greisenized rocks in a granite roof zone, such as the Haad Som Pan deposit in Thailand (Aranyakanon, 1961). These are not thought to be disseminations in a magmatic sense and the complexity of the pneumatolytic alteration processes at Haad Som Pan has been emphasized by Aranyakanon (1961). (c) The common occurrence of stanniferous pegmatites and aplites intruding both granite and country rock and found especially in the roof zones of intrusions. Although

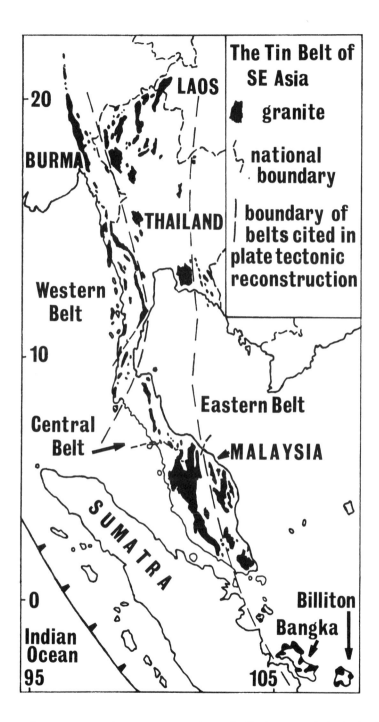

The Tin Belt of SE Asia

▪ granite

↳ national boundary

⎮ boundary of belts cited in plate tectonic reconstruction

LAOS

BURMA

THAILAND

Western Belt

Eastern Belt

Central Belt →

MALAYSIA

SUMATRA

Indian Ocean

Billiton
Bangka

20

10

0

95

105

Figure 1 Generalized map showing the approximate distribution of granitoids in the tin belt of SE Asia. The zone of subduction of oceanic lithosphere beneath Sumatra is indicated in the bottom left corner

the latter are typically thin intrusions ($<$ 2 m) they are often quite coarse-grained, and frequently carry cassiterite which is visible to the naked eye. Hosking (1969) has objected to the use of the term 'pegmatite' for these bodies. However, although there may be a complete gradation between 'veins' and 'pegmatites', in the present writer's experience, the texture and mineralogy typical of these late intrusions, which contain K-feldspar, albite, quartz, tourmaline, garnet and cassiterite, often makes the term 'pegmatite' a natural choice in the field.

Although it is clear that the primary tin mineralization is spatially associated with particular granites, the *precise* geochemical relationship between the granite and tin mineralization is not well understood. In general, it has not been established that particular cassiterite-bearing 'pegmatites' which intrude a granite actually fractionated from the same magma that produced the granite host rock. Similarly, the source of the fluids involved in the formation of hydrothermal vein swarms or in the pneumatolytic alteration which produced the Haad Som Pan type of disseminated deposit is generally unknown. By analogy with similar ore deposits elsewhere, it would be reasonable to suspect that meteoric water might be involved (Taylor, 1974) and D/H analyses on muscovites from the Mae Lama hydrothermal vein system in North-west Thailand have proved that this is the case for this particular primary tungsten-tin deposit (Harmon and Beckinsale, unpublished). This raises a secondary problem — whether the interaction with meteoric water is perhaps only an extra stage in concentrating ore metals derived from the granite, or whether the meteoric water that gained access to the cooling pluton contained dissolved ore components before interacting with a granite which acted only as a heat source to drive the hydrothermal circulation. Finally, in both eastern Malaysia and Billiton, Indonesia, there are mined occurrences of primary tin mineralization of major economic significance which consist of bedded cassiterite–pyrrhotite–magnetite ores interstratified with sediments. These ores have traditionally been considered to be of pyrometasomatic origin (Adam, 1960; Bean, 1969; Hosking, 1973), similar to known skarn deposits containing both cassiterite and malayaite ($CaO.SnO_2.SiO_2$) such as that at Pinyok Mine, Thailand (Hosking, 1969) and related in some way to granites known to be of Triassic age (Bignell and Snelling, 1977; Priem *et al.*, 1975). However, if a recent suggestion (Hutchison and Taylor, 1978) that some of these bedded ores may be of volcanogenic/exhalative origin proves to be correct, then the ores themselves must be older than any of the dated granites which intrude them.

Despite these areas of uncertainty in the precise geochemical relationships between primary tin mineralization and granite magmatism, it is evident from field experience that tin deposits tend to occur in association with granites with highly evolved geochemical characteristics (Pitakpaivan, 1969). Moreover, the most typical SE Asian tin-granite would be classified as an 'S-type' granite (Chappell and White, 1974) or 'ilmenite-series' granite (Ishihara, 1977) rather than an 'I-type' or 'magnetite-series' granite (Beckinsale *et al.*, 1979). Thus, in order to pursue the geochemical evolution of the SE Asian tin belt back one stage in the sequence — *Quaternary placers* derived from *primary tin mineralization* associated with *granites* — the next section will be devoted to the classification of granite magma types in relation to their plate tectonic settings and to the petrogenesis of 'S-type' and 'I-type' granites.

Classification of granite magma types in relation to plate tectonic setting

The classification of granites into 'I-types' or 'S-types' was proposed by Chappell and White (1974) and based on studies of granites forming major batholiths of the Tasman Orogenic Zone of eastern Australia. This classification is approximately equivalent to the distinction between 'magnetite-series' granites and 'ilmenite-series' granites, respectively, proposed by Ishihara (1977 and earlier papers). The geochemical and mineralogical distinctions between these granite types which the present writer considers the most important tendencies are summarized in Table 1.

Table 1 **Characteristic features of I- and S-type granites**

I-type or magnetite-series granites	S-type or ilmenite-series granites
Tend to be the acid end of a broad compositional spectrum from basic to acid	Tend to occur in restricted ranges of only acidic compositions
Relatively high sodium contents	Relatively low sodium contents ($< 3.2\%$ Na_2O in rocks with $\sim 5\%$ K_2O)
Low initial $^{87}Sr/^{86}Sr$ ratios (<0.708)	High initial $^{87}Sr/^{86}Sr$ ratios (> 0.708)
Normal range of $\delta^{18}O$ values (approx. $6-10\%_0$, SMOW)	Enriched in ^{18}O ($\delta^{18}O$ values $\geqslant$ about $10\%_0$, SMOW)
Magmas with relatively high oxygen fugacity; relatively high ferric/ferrous ratios; characterized by magnetite	Magmas with relatively low oxygen fugacity; relatively low ferric/ferrous ratios; characterized by ilmenite
Hornblende and sphene commonly present	Muscovite, monazite, cordierite and garnet commonly present

As with every other classification of igneous rocks, none of the distinctions listed in Table 1 should be regarded as absolutely rigid. Note that Coleman (this volume) has demonstrated that in some cases significant changes in oxygen fugacity can occur during crystallization of a granite magma, and in such a case the distinction between I-type and S-type granite based on the presence of magnetite or ilmenite could be misleading. Nevertheless, the tendencies indicated in Table 1 are thought to reflect fundamental differences in the source regions of the two types of granite. I-types are derived from igneous material whereas S-types are derived from a source region within the continental crust. For example, the low sodium content of S-type granites is thought to reflect the involvement of the source region in a previous sedimentary cycle with concomitant partition of sodium into seawater. Similarly, the relatively low oxygen fugacity characteristic of S-type magmas is thought to reflect the presence of reducing agents such graphitic shales in the continental crust. The high initial $^{87}Sr/^{86}Sr$ ratios and ^{18}O-enriched oxygen isotope compositions (O'Neil and Chappell, 1977) in the S-type granites reflect the relatively high bulk Rb/Sr ratios and high $\delta^{18}O$ values of continental crust in comparison with the mantle. The fundamental difference is between fractionation of a mantle-derived basaltic parent (possibly via more than one melting event) leading to an I-type granite, and fractionation over a more limited range of silica contents to produce various S-type granites from crustal melts. It should be noted that if a mantle-derived volcanic complex which had become part of the continental crust were to be remelted, the resulting partial melts would be characterized by I-type chemistry and, in this case, the distinction between a mantle-derived magma and a crustal-derived magma would be difficult to establish.

Both I-type and S-type magmas experience fractionation leading to more evolved compositions, and both magma types may be contaminated before, during or after fractionation. It is therefore necessary to be cautious when using the criteria listed in Table 1 to distinguish between the two types of granite. Provided the criteria in Table 1 are applied with restraint, however, they are useful in a surprising number of geological situations. In the case of the granites in the Tertiary volcanic complex of the Isle of Mull, NW Scotland (Walsh *et al.*, in press), for example, it has been established that the early (Centre 1) granites are predominantly crustal melts, although they are contaminated with basaltic fractionation products, and show some of the characteristics of S-type granites in that Na_2O contents are low (averaging about 3.0%) and initial $^{87}Sr/^{86}Sr$ ratios are high (averaging about 0.715). The later granites of Centre 2 and Centre 3 are thought to be predominantly products of fractionation of a basaltic parent with some component of crustal contamination. The latter granites are part of a spectrum of compositions from basic through intermediate to acid rocks. They have high Na_2O contents (averaging about 4.0%) and relatively low initial $^{87}Sr/^{86}Sr$ ratios in the range 0.705–0.710, and thus show many I-type characteristics. In the Tertiary volcanic complex of Mull, however, *all* the granites are strongly depleted in ^{18}O as a result of late-stage interaction with meteoric groundwater, and the isotopic composition of oxygen is not a useful criterion for distinguishing granite types in this case.

At this point it is useful to examine Figure 2 which shows the distribution of I-type and S-type granites on a world scale. Although Figure 2 is very much simplified, it is immediately apparent that many batholithic granites in which I-types predominate are spatially related to present-day subduction of oceanic lithosphere. Good examples are the Andean batholiths related to the subduction of the Pacific oceanic plate beneath the South American continent and the intrusives of the Philippines related to eastward subduction of the oceanic lithosphere of the South China Sea *and* westward subduction of the oceanic lithosphere of the Philippine Sea. Note also that porphyry copper deposits are characteristically associated with I-type, subduction-related granites. For example, a large number of porphyry copper deposits are known in both the Philippines and certain parts of the Andes. No doubt examples of S-type granites may be found in both the

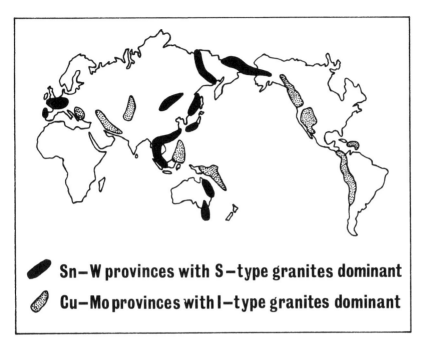

Figure 2   Map showing distribution of I-type and S-type granites on a world scale, based on the work of Ishihara (1977)

Andes and the Philippines (related, as in the case of Mull, to local fusion of continental crust), but the predominant chemistry of the granites in both areas is I-type, reflecting a mantle source region (see Atherton *et al.,* this volume). In contrast most of the batholithic granites in which S-types predominate and which are characteristically associated with tin deposits are *not* obviously spatially related to subduction of oceanic lithosphere at the present time. The only situation in current plate tectonic models, other than subduction, in which it would be possible to produce a batholithic scale belt of S-type granites via extensive crustal melting, is that of collision between continents or between a continent and a magmatic arc. It is important to remember, however, that such collisions only occur when an ocean or marginal basin closes, and closure can only occur by subduction of oceanic lithosphere. Thus, we would expect to find a belt of older I-type granites, related to earlier subduction of oceanic lithosphere, on the continent side of a younger belt of S-type granites.

This simple model can be applied to the geological history of the tin belt of SE Asia. The basic tenets of the model are: (a) belts of I-type granites and associated porphyry copper deposits are related to subduction of oceanic lithosphere; (b) belts of S-type granites with associated tin deposits are related to continental collisions with an older belt of I-type granites on the continent side of them. In order to complete the reconstruction it is necessary to know the age of the granites, hence aspects of the geochronology of the tin belt are discussed in the next section.

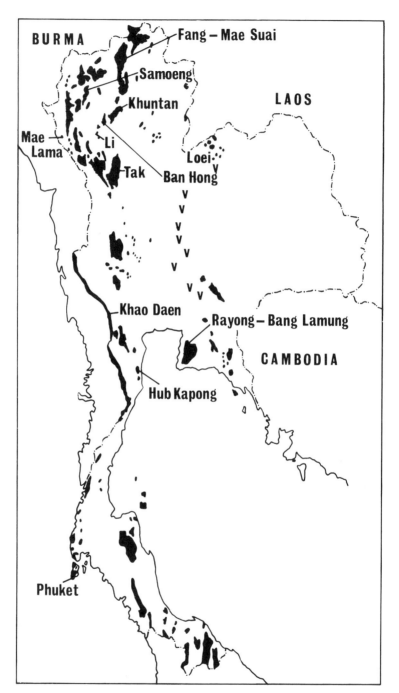

Figure 3 Map showing distribution of granites in Thailand. The granites for which there are available Rb-Sr whole-rock isochron ages (Table 2) are named. Permo-Triassic volcanic rocks forming part of the Eastern Belt magmatic arc are indicated by the letter V

Geochronology and a proposed plate tectonic reconstruction of the tin belt of SE Asia

Although more good-quality geochronological data are needed for the SE Asian tin belt, a substantial number of results are available (e.g. Priem *et al.*, 1975; Bignell and Snelling, 1977; Beckinsale *et al.*, 1979). Some general problems in the interpretation of geochronological data for tin belt granites have come to light, the most important being that K-Ar ages for separated micas are often grossly discordant from ages derived from statistically acceptable Rb-Sr whole-rock isochrons. It has been concluded that generally in this area the Rb-Sr whole-rock isochron ages approximate to the age of intrusion of the granite and that the K-Ar ages have been reset. Thus, it is essential to rely only on Rb-Sr whole-rock isochron ages in attempting a plate tectonic reconstruction and to ignore K-Ar age data. In the following discussion attention will be focussed on Thailand because it is the area within the tin belt with which the writer is most familiar. Furthermore, detailed studies of the major and trace element geochemistry of most of the dated granites in Thailand are available to enable classification into I- or S-types (Teggin, 1975; Pongsapich and Mahawat, 1977; Suensilpong *et al.*, 1977; Beckinsale *et al.*, 1979 and unpublished data). The distribution of granites

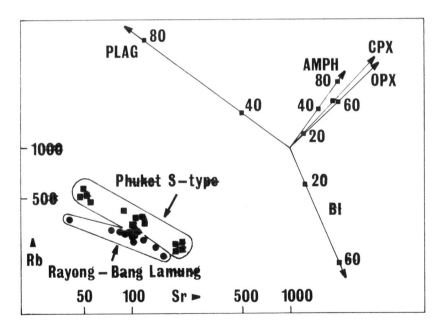

Figure 4 A plot of Rb versus Sr for two typical Thai granites. The vectors indicate the change in composition of a melt as a result of fractional crystallization of the named phenocryst phases in acid rocks. The percentage of fractional crystallization of the original melt to produce corresponding changes in Rb and Sr is marked along the vectors. PLAG = plagioclase; AMPH = amphibole; CPX and OPX = ortho- and clinopyroxene; BI = biotite.

in Thailand is shown in Figure 3 and the available Rb-Sr data are summarized in Table 2 and Figure 4. Although S-type granites clearly predominate, I-types are also present (Table 2) and in all cases except one the classification is clear-cut. The one exception is the Tak batholith which has features of both I- and S-type granites and is thus perhaps of mixed origin. The plot of Rb versus Sr (Figure 4) shows that, within the individual intrusions yielding Rb-Sr whole-rock isochrons, the range in chemical composition is probably related to fractional crystallization of plagioclase, and that quite extensive fractionation is required to produce the range in Rb-Sr ratios. However, it should also be noted that the granites associated with the richest tin mineralization have the highest initial $^{87}Sr/^{86}Sr$ ratios, above

Table 2 **Geochronological data for Thai granites**

Location (Figure 2)	Rb-Sr whole-rock isochron age and 2σ error Ma	Initial $^{87}Sr/^{86}Sr$ and 2σ error	Classification	Reference
Khao Daen	93 ± 4	0.7338 ± 7	S-type	Beckinsale *et al.*, 1979
Phuket area	108 ± 5	0.7293 ± 5	S-type	Beckinsale, unpublished
Phuket area	124 ± 4	0.7073 ± 13	I-type	Snelling *et al.*, 1970
Mae Lama	130 ± 4	0.7086 ± 7	I-type	Beckinsale *et al.* 1979
Samoeng	204 ± 15	0.7328 ± 21	S-type	Teggin, 1975
Hub Kapong	210 ± 4	0.7237 ± 6	S-type	Beckinsale *et al.*, 1979
Khuntan	212 ± 12	0.7244 ± 20	S-type	Teggin, 1975
Tak (white)	213 ± 10	0.7158 ± 13	?	Teggin, 1975
Tak (pink)	219 ± 12	0.7104 ± 19	?	Teggin, 1975
Rayong-Bang Lamung	220 ± 13	0.7265 ± 13	S-type	Beckinsale, unpublished
Fang-Mai Suai	240 ± 64	0.7280 ± 66	S-type	Von Braun *et al.*, 1976
Ban Hong	242 ± 9	0.7253 ± 15	S-type	Von Braun *et al.*, 1976
Li	244 ± 28	0.7220 ± 44	S-type	Von Braun *et al.*, 1976

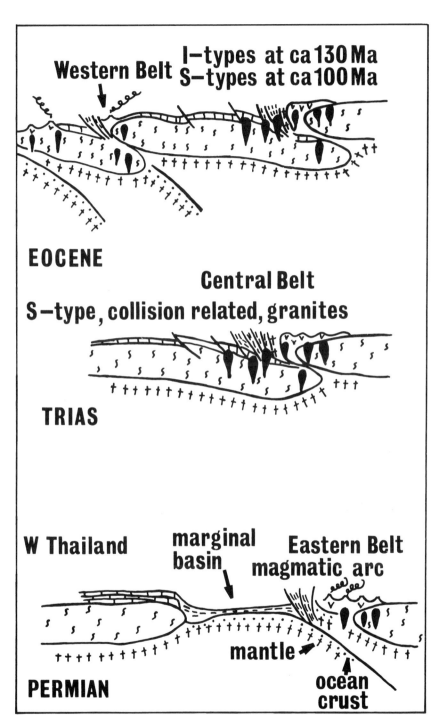

Figure 5 Cartoon illustrating a plate tectonic reconstruction for the tin belt of SE Asia based on the work of Mitchell (1977) and discussed in the text

0.728. The Khao Daen, Samoeng and Phuket S-type granites all have current mining activity and it is possible that tin was concentrated in these intrusions, together with Rb and other incompatible elements, by more than one partial melting – fractional crystallization cycle.

The geochronological and geochemical data for Thai granites summarized above is in very good agreement with a plate tectonic reconstruction proposed by Mitchell (1977) using other lines of evidence, although some minor modification of his model is required (see Figure 5). In Permian times an ocean or, more probably, a marginal basin occupied the area of present central Thailand, and eastward subduction of oceanic lithosphere is proposed to account for the Permo-Triassic volcanic arc (Figure 3), the known porphyry copper deposit at Loei and the granites of the Eastern Belt. The Eastern Belt granites of Thailand outcrop along strike from the East Coast Belt of Malaysia, which Hutchison and Taylor (1978) also identify as a 'slightly eroded volcano-plutonic arc'. The marginal basin closed with a continental collision in Triassic times (about 240–210 Ma) giving the Central Belt of predominantly S-type granites forming the main tin belt in Thailand (which may be correlated along strike with the Main Range granites in Malaysia). In Lower Cretaceous times (about 130 Ma) I-type granites were emplaced in West Thailand at Mae Lama and Phuket. Again, it is proposed that these granites reflect eastward subduction of oceanic lithosphere probably from a marginal basin lying to the west of Thailand, before Burma drifted into its present position. This marginal basin also closed with a continental collision in roughly Middle Cretaceous times (about 100 Ma) producing the S-type granites at Khao Daen and at Phuket, both of which are richly mineralized. The small scale of plate tectonics in this reconstruction is consistent with the present situation in SE Asia and is very different from the large scale and constancy of the situation in the Andes, for example, where it would appear that subduction of oceanic lithosphere has gone on in one direction for about 250 Ma. South-east Asia occupies a position between the major Eurasian, Pacific and Indian Ocean plates, and has been characterized by opening and closure of marginal basins, rapid changes in the pattern of subduction, and collisions between continents (or microplates) or between a continental plate and a magmatic arc about every 20 Ma.

In conclusion, the model presented above for the tin belt of SE Asia is that tin is generally concentrated in continental crust and primary tin mineralization may develop when crustal rocks experience a partial melting – fractional crystallization cycle – (perhaps more than once) producing S-type granites with highly evolved geochemical features. The plate tectonic setting in which bulk melting of continental crust has occurred, leading to batholithic scale belts of S-type granites, is in collisions between continents, or between continents and magmatic arcs. One remaining problem is the situation in Japan where young, S-type granites with tungsten and tin mineralization are found on the ocean side of an island arc (Ishihara, 1978). Perhaps recycling of older crustal material down a Benioff Zone is the most reasonable suggestion to account for such occurrences.

Published with permission of Director, Institute of Geological Sciences.

44

THE GEOCHEMICAL CHARACTER OF THE SEGMENTED PERUVIAN COASTAL BATHOLITH AND ASSOCIATED VOLCANICS

M.P. ATHERTON, W.J. McCOURT, L.M. SANDERSON and W.P. TAYLOR
The Jane Herdman Laboratories of Geology, University of Liverpool,
Brownlow Street, P.O. Box 147, Liverpool L69 3BX, UK

Introduction

The Peruvian Coastal Batholith and the associated volcanics and sediments have been studied in some considerable detail by workers from the University of Liverpool and the Institute of Geological Sciences (see Pitcher, 1978). These studies have been predominantly concerned with the detailed mapping of the batholith, its stratigraphic and structural setting, and with the petrographic variation.

On the basis of these studies, Cobbing *et al.* (1977 b) divide the batholith into three major segments (from north to south): the Trujillo, Lima and Arequipa segments. Each of these is composite in nature, consisting of numerous separated plutons, dykes and sills assembled into major plutonic complexes. Thus, there are some 230 individual plutons in the Lima segment and perhaps 1000 in the whole batholith, which stretches some 1600 km, parallel to the trench. The nature of this 'plutonic lineament' has been emphasized by Pitcher (1978), who suggested that magma intrusion was focussed up a linear structure throughout the life of the batholith ($\sim$70 Ma), to freeze at a subvolcanic level just short of the surface. In Peru, the plutonic rocks intrude Mesozoic-Cenozoic sediments and volcanic 'arcs' which lie on thick continental crust (50–70 km) mainly Precambrian in age ($\sim$1900 Ma, Cobbing *et al.*,1977b; Ries, 1977).

The detailed work on the Peruvian Coastal Batholith makes it a prime subject for geochemical investigations leading to an understanding of the processes of batholith generation. Gill (1978) has stressed, with reference to trace element models in studies of andesite genesis, that 'future trace element studies need to be accompanied by more thorough field, petrographic, and isotope work than has been common heretofore' (Gill, 1978, p. 721). Even with some of these conditions nearly fulfilled, as in Peru, there is still a need for continuing inter-action between fieldwork and geochemical studies in order to eliminate and constrain some of the possible models.

Geological setting and components of the Peruvian Batholith

The major components of the batholith and associated rocks (Figures 1 and 2), which are the concern of this paper, are:

45

1. Country rocks: mainly Upper Cretaceous volcanics (Casma), with some Jurassic volcaniclastics and Lower Cretaceous sediments in the south.
2. Cover rocks: mainly Eocene-Miocene volcanics (Calipuy) in the north plus Recent volcanics in the south (Cobbing *et al.,* in press).
3. Plutonic rocks intruding (1):

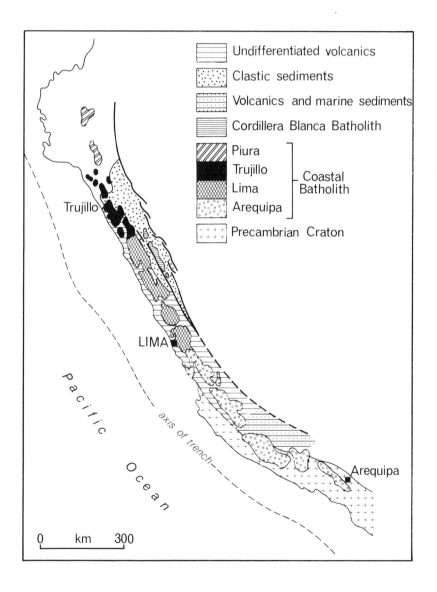

Figure 1 The segmented Coastal Batholith of Peru and its geological setting (after Cobbing, 1978); details of country and cover rocks given in the text

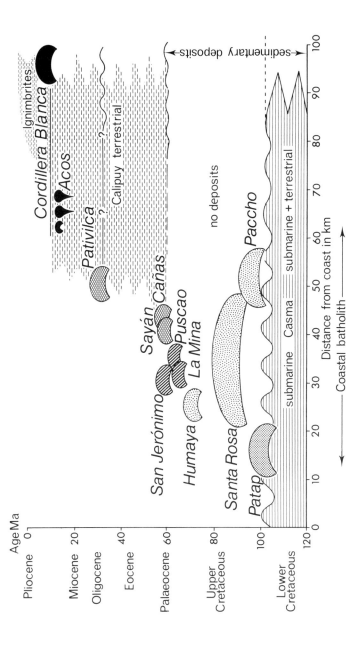

Figure 2 Temporal and spatial relationship of plutonism, volcanism and tectonism in the Western Cordillera of North Peru (after Pitcher, 1978). Ages of plutonics (mainly K–Ar) given by bases of symbols. Close dots = gabbro; open dots = early granodiorites; diagonal lines = granitoids of ring complexes; cross-hatched = late granite plutons, of which Sayán and Pativilca have K-feldspar phenocrysts; black = granites and granitoids of the Cordillera Blanca and eastern stocks, respectively

47

(a) Gabbros and diorites of the Patap Superunit, often showing evidence of initial cumulate crystallization at depths of up to 35 km (Regan, 1976) followed by further crystallization near or at the present crustal level. Many of these rocks have had a long, complex history of crystallization involving crystal growth and accumulation, metamorphism, metasomatism, deformation and hybridization (Regan, 1976). They carry plagioclase ± olivine ± clinopyroxene ± orthopyroxene ± hornblende as primary phases, the hornblende becoming a major phase during late-stage crystallization as P_{H_2O} increased (Mullen and Bussell, 1977).

(b) An earlier group of granitoids, which is made up of superunits varying from quartz diorite to granite in composition, with a weighted average near to granodiorite (Figure 3). The Santa Rosa superunit of the Lima segment may be considered as an example of this group and, although the mean compositions of other superunits may be more basic or acidic, they all show a basic marginal facies and a crude increase in acidity towards a centre, which is often a late mobile phase cross-cutting earlier members. Individual units of a superunit are often confined within mappable contacts and represent upward 'surges' in the same magma chamber (Pitcher, 1978). The crystallization of these superunits involved pyroxene, hornblende, plagioclase, K-feldspar, biotite and quartz, often with pyroxene or hornblende dominating as the main mafic phase, depending on the physical conditions during crystallization of the superunit. The textures and general 'zonal' character of the plutons indicate that crystallization was mainly at the present level, although plagioclase cores of An_{80} and mafic clots demonstrate the importance of an earlier imprint which may be that of the source region (White and Chappell, 1977). Indeed, all these superunits have typical 'I' type characteristics. Interestingly, in the south, the crystallization of the superunits is more 'telescoped' and acid end-members are rare compared to the north (Figure 3). Whatever the exact mechanism causing the zoning, it is clear that the crystallization front moved in from the margins of the plutons and that liquid separation was never complete (cf. Bateman and Chappell, 1979). Typically, the superunits bear few pegmatites and show little evidence of re-equilibration, hence the magmas appear to have been undersaturated with water for most, if not all, of their crystallization history. However, fine-grained equigranular groundmass textures in some parts of the superunits are, we believe, quenched textures due to rapid subsurface venting of fluid probably via fractures above the magma chamber. This is compatible with intrusion of these magmas near to the surface (Atherton and Brenchley, 1972; Taylor, 1976), and the close association with the ring dyke complexes (Myers, 1975). Bussell et al. (1976) indicate that such complexes were probably present, albeit in a primitive form, at an early stage in the history of the batholith.

(c) A later group of superunits, all of similar age (Figures 2 and 3). These form major ring complexes and are separated in time from the earlier granitoids by a major dyke swarm. Some show cryptic zoning, e.g. the Puscao unit (Taylor, 1976), and others exhibit multipulse behaviour which operated during cauldron subsidence. The latter may possibly be the more mobile phases of deeper differentiated magma chambers and, as such, tend to be fairly silicic. Many of these rocks

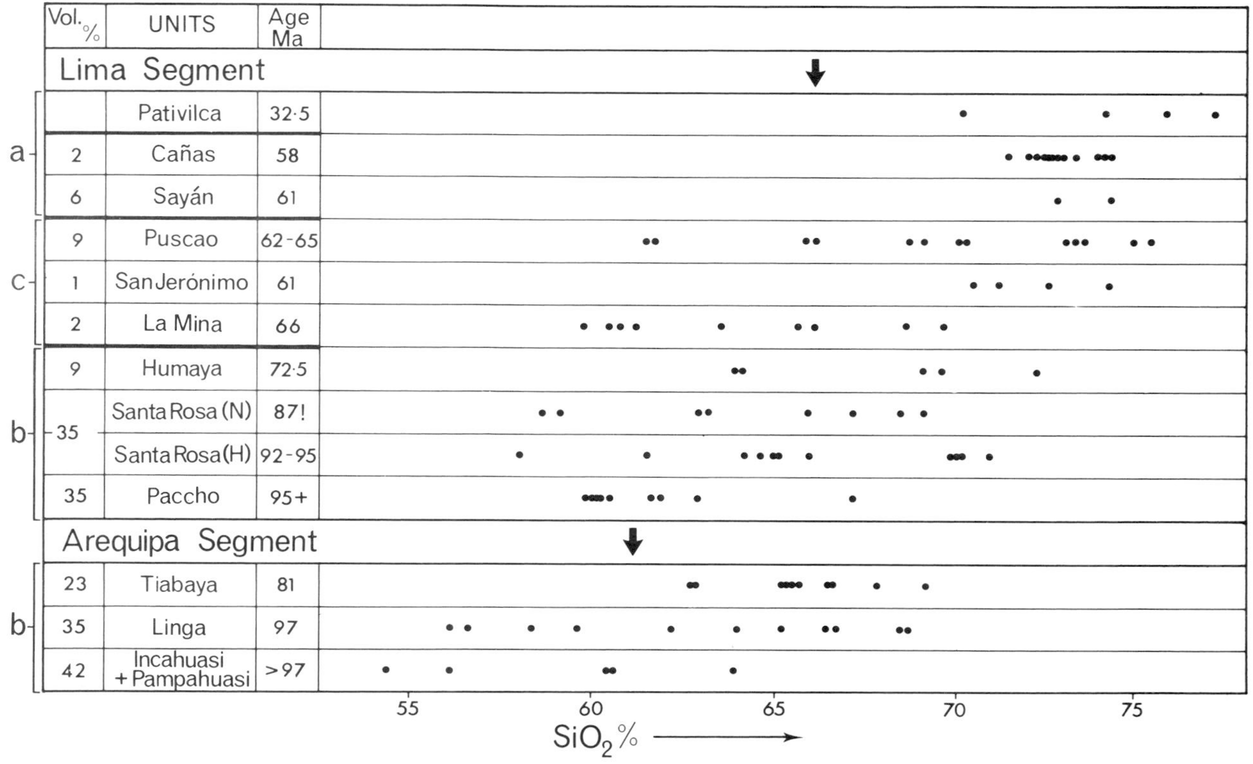

Vol.%	UNITS	Age Ma	SiO₂ chart

Lima Segment

Vol.%	UNITS	Age Ma
	Pativilca	32·5
2	Cañas	58
6	Sayán	61
9	Puscao	62-65
1	SanJerónimo	61
2	La Mina	66
9	Humaya	72·5
35	SantaRosa (N)	87!
	SantaRosa(H)	92-95
35	Paccho	95+

Arequipa Segment

Vol.%	UNITS	Age Ma
23	Tiabaya	81
35	Linga	97
42	Incahuasi +Pampahuasi	>97

SiO_2 % ⟶

55   60   65   70   75

Figure 3   Silica contents of superunits and plutons of the Lima and Arequipa segments of the Coastal Batholith, showing the general increase in acidity with decreasing age (K–Ar dates). Arrows indicate weighted SiO_2 average of each segment

49

show effects of degassing, such as granophyric, vesicular, tuffisitic and explosion textures, as well as decreasing grain size upwards (Bussell and McCourt, 1977). These are the rocks classically defined in the four ring complexes of the Lima segment which Bussell *et al.* (1976), Myers (1975), Pitcher (1978) and others have interpreted as 'representing a direct connection between the intrusives of the batholith and the volcanics of the country rocks and cover' (Pitcher, 1978); thus, 'the plutons rose into their own ejecta' (Myers, 1975).

(d) A final group of separated single plutons, which post-date the main ring dyke event and includes the Pativilca, Sayán and Cañas intrusions. They are all granitic in composition and may show a pegmatitic facies. Cañas clearly differentiated upwards in a sealed chamber to produce a high-level cryptically zoned pluton (Taylor, 1976). However, Pativilca and Sayán have petrographic characteristics, such as K-feldspar phenocrysts and large grain size, which may place them outside the main evolutionary trend of the batholith, and probably Pativilica is much younger (Figure 2).

The main elements of the batholith and associated rocks outlined above must, we believe, be integrated in any coherent account of batholith genesis. However, there are weaknesses in our understanding of these elements, which in particular concern the volcanics. First, the stratigraphy of the Casma and particularly the Calipuy volcanic groups is, except for a few worked sections, poorly understood. This is partly due to the formidable thicknesses and extent of the volcanics, but also to the great change in thickness of the Casma volcanic sequence across the axis of the batholith and to the rapid lateral variations in both sequences. Determining which are the basal members of the Calipuy volcanics or recognizing lateral equivalents within the sequence is, therefore, difficult. Alteration of the volcanics by burial metamorphism (Webb, 1976; Aguirre *et al.,* 1978) and hydrothermal activity is common and makes absolute age determination and chemical modelling difficult. K–Ar age studies of the plutons (Wilson, 1975; Moore, in preparation) have also indicated that problems exist with respect to the local resetting of the ages of the younger ring complex members in the north and the more comprehensive resetting of the superunits in the south.

These weaknesses are crucially important to appreciate in any model relating plutonics to volcanics, or relating pluton sequences along the batholith axis. However, we can put forward the important problems to be solved in a coherent batholith model, bearing in mind the weaknesses outlined above and the field-based nature of the sampling.

Problems

SUPERUNITS

Superunits have been recognized in the Southern California Batholith (Larsen, 1948), the Sierra Nevada Batholith (Bateman and Dodge, 1970), and in Peru (Cobbing and Pitcher, 1972), and are considered to be major assemblages of

Table 1 Chemical analyses of representative samples of the Peruvian Coastal Batholith and averages of associated rocks

Ref.	Gabbros A	Volcanics B	C	D	E	Stocks F	Coastal Batholith G	H	I	J	K	L	M	N
No. of analyses	8	10	4	7	5	10	1	1	1	1	1	1	1	1
SiO_2	48.29	52.23	71.34	57.62	69.89	64.50	59.08	66.19	57.24	63.86	58.53	69.45	71.35	75.34
TiO_2	0.62	0.74	0.25	0.88	0.34	0.59	0.64	0.62	0.88	0.66	0.99	0.36	0.32	0.18
Al_2O_3	19.07	17.70	13.96	19.57	14.35	15.77	17.08	15.52	17.96	16.80	16.25	14.88	14.32	12.87
Fe_2O_3	2.78	3.08	0.98	3.22	1.13	1.36	2.37	1.74	2.43	1.07	2.94	1.59	2.00	0.25
FeO	6.98	6.72	1.52	4.05	1.83	1.30	4.98	2.59	5.47	4.19	5.69	1.23	0.65	0.80
MnO	0.18	0.18	0.06	0.15	0.08	0.06	0.15	0.08	0.18	0.10	0.18	0.07	0.09	0.09
MgO	7.96	5.08	0.48	2.76	1.05	1.51	2.72	1.77	2.82	1.60	3.07	1.24	0.97	0.41
CaO	11.20	10.21	2.55	6.60	2.57	3.93	5.70	3.11	5.66	4.28	6.22	2.81	2.26	0.81
Na_2O	1.97	2.59	2.91	2.84	3.79	3.52	3.07	2.82	3.36	3.00	3.09	3.69	3.71	3.88
K_2O	0.24	0.55	4.20	1.65	3.49	3.33	2.64	4.42	1.78	3.67	1.57	3.29	3.13	4.38
P_2O_5	0.12	0.15	0.04	0.19	0.07	0.22	0.17	0.14	0.20	0.30	0.15	0.05	0.06	0.03
Total	99.41	99.23	98.29	99.53	98.59	96.09	98.60	99.00	97.98	99.53	98.68	98.66	98.86	99.04
Co	49	40	—	15	—	22	16	12	17	13	19	7	5	2
Cr	18	11	—	13	—	9	15	2	13	nd	5	10	3	16
Ni	27	23	—	3	—	13	5	10	10	7	8	23	13	11
Rb	6	21	219	33	135	100	118	261	71	165	59	99	122	103
Sr	427	255	107	368	128	603	517	282	428	360	328	306	238	82
V	271	262	—	168	—	—	142	86	180	69	181	47	39	12
Y	8	17	35	26	40	79	26	23	35	41	43	11	14	14
Zr	27	41	211	148	259	154	137	277	154	261	129	113	96	87

(Oxides in weight percent; trace elements in ppm; nd = not detected; — = not analysed.)

A = Primary gabbros.
B = Basic Casma volcanics.
C = Acid Casma volcanics.
D = Basic Calipuy volcanics.
E = Acid Calipuy volcanics.

F = High level stocks (Harding, 1978).
G = Linga superunit, Arequipa segment.
H = Linga superunit, Arequipa segment.
I = Incahuasi superunit, Arequipa segment.
J = Incahuasi superunit, Arequipa segment.

K = Santa Rosa superunit, Lima segment.
L = Santa Rosa superunit, Lima segment.
M = Puscao superunit, Lima segment.
N = Pativilca pluton, Lima segment.

51

consanguineous rock types. They may be recognized in Peru as a number of uniquely identifiable units, usually ranging from quartz diorite to leuco-monzogranite in composition, which show, often over long distances, 'consistent spatial and temporal association of all the constituent units, . . . progressive evolution of both the mineralogy and the textural interrelations, especially of quartz and K-feldspar and the maintenance of similar granularity and aphyric character' (Cobbing *et al.*, 1977 b). The Coastal Batholith is made up of relatively few of these superunits, 8 or 9 in the north and 5 in the south. A unique age and composition is implied by these authors, and follows from the original mapping of such 'lithostratigraphic' units.

The chemistry of the major elements indicates that the variation within the main superunits is similar and due to 'high-level' differentiation (Figure 5; Table 1). Thus they all follow a common calc-alkaline trend and plot with similar trends on Harker diagrams. This conclusion is confirmed by both the petrology and the detailed geochemical studies of Taylor (1976), which support a model of differentiation at or near the present level to account for the variation seen. However, these data do *not* indicate that the rocks are from a common parent magma (Le Maitre, 1976). Indeed, two superunits, La Mina and Santa Rosa, have almost identical compositional spreads on AFM and Harker plots, but are separated by 30 Ma in age. Furthermore, on an Ab-An-Or-Qz $(-H_2O)$ diagram (Figure 6), each superunit defines its own smooth curve originating from different points in the plagioclase field, at a high angle to the quartz saturation surface. This is indicative of separate equilibrium fusion events in the source area for each superunit (Presnall and Bateman, 1973; McCourt, 1978).

The trace element geochemistry supports this model, thus Rb and Rb/Sr increase with fractionation while Sr, K/Rb and Ca/Y all decrease. In all cases the trace element data define separate trends for each superunit (Figure 7; McCourt, 1978), and this agrees with a model of a separate equilibrium fusion event followed by high-level fractionation for each superunit.

An interesting but detailed point with regard to the superunit concept concerns the Santa Rosa superunit which has been studied in detail in two traverses some 300 km apart (Atherton and McCourt, in preparation). On the basis of the arguments expressed above, the Santa Rosa evolved from two distinct magmas which intriguingly encompass the total compositional variation seen in all the super-units (Figures 7 and 8). This challenges the concept of the superunit in the sense of an implied consanguinuity, and has important consequences for petrogenetic modelling of granitoid rocks. Accepting the above, we have at least 8 separate magma batches forming the batholith in the Lima segment and 5 in the Arequipa segment.

The general high-level evolution of the superunits can be modelled in terms of plagioclase ± pyroxene ± hornblende fractionation. The Rb and K/Rb variations (Figure 7) are compatible with plagioclase and pyroxene as important fractionating phases, while Y and rare earth element (REE) trends (Figure 4) indicate that pyroxene was succeeded by hornblende at tonalitic compositions in some super-units, with probably sphene and/or apatite becoming important in the late 'granitic' superunits. Thus, plagioclase plus pyroxene and/or hornblende dominated fractio-

52

nation accounts for the calc-alkaline character of the individual superunits and hence of the batholith (Figure 5).

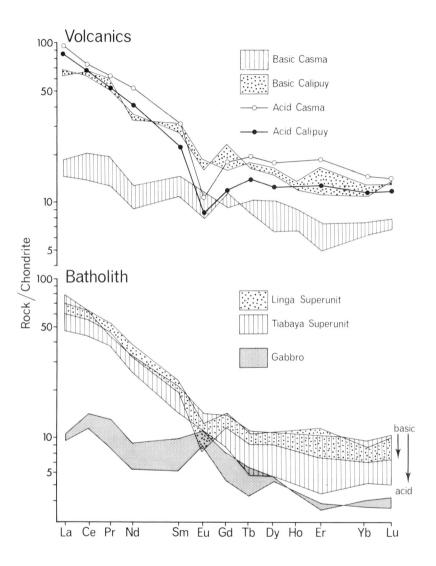

Figure 4 REE abundances in samples of the Coastal Batholith and associated rocks normalized against the Leedey chondrite (Masuda *et al.*, 1973). Shaded areas represent ranges in composition. Arrows for Linga and Arequipa superunits indicate variation with increasing SiO_2 content

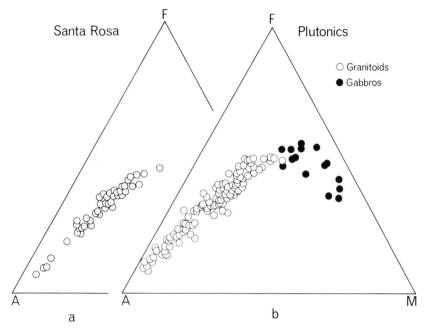

Figure 5 (a) AFM diagram of compositions of rocks of the Santa Rosa superunit, Lima segment, Coastal Batholith. (b) AFM diagram of compositions of gabbros and granitoids of the Trujillo, Lima and Arequipa segments, Coastal Batholith

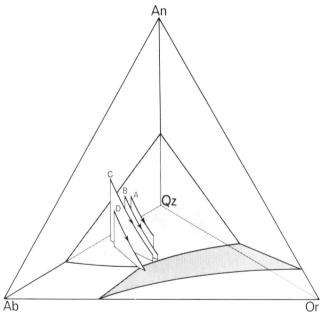

Figure 6 An–Ab–Or–Qz(–H$_2$O) diagram showing the separate crystallization paths of the major superunits of the Lima segment of the Coastal Batholith. A = Santa Rosa (Nepeña); B = rocks of the Huaura complex; C = Santa Rosa (Huaura); D = San Jerónimo

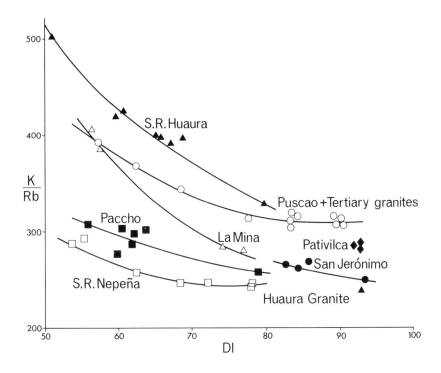

Figure 7 Plot of K/Rb against differentiation index for rocks of the Lima segment, Coastal Batholith

SEGMENTS

Mapping of the batholith to the south of Lima indicated that the superunits change, apart from the Patap (Gabbro), and reduce to 4 in number (Tiabaya, Linga, Incahuasi and Pampahuasi), stretching some 900 km south of Lima. These specific assemblages of superunits define different segments of the batholith. There are also two, as yet poorly defined, segments in the extreme north, namely the Trujillo and Piura. We will consider only the Lima and Arequipa segments here, our concern being the differences in the segments and their significance.

The major differences between the two segments are outlined in Table 2, and one immediately obvious difference relates to water—rock interaction. Specifically, the primary igneous minerals in the Arequipa segment are often altered to chlorite, sericite and epidote, while late K-feldspar mobilization with marked reddening of the rocks, particularly in the Linga superunit, is common. Some units have specific alteration patterns (Agar, 1978) and there is a clear relationship between mineralization and plutonism, such that, locally, individual superunits have specific mineralization associated with them, e.g. Linga — widespread and low-grade porphyry copper (Agar, 1978). Finally, there is comprehensive resetting of the older K—Ar ages by the youngest superunit of this segment (Moore, personal communication), which is a clear indication of late hydrothermal activity (Taylor, 1977). The effects of the water—rock interaction are striking and make the rocks

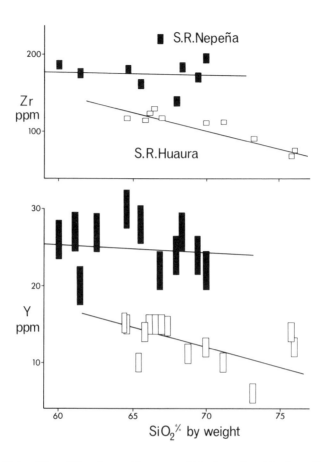

Figure 8 Plots of Zr and Y against weight percent SiO$_2$ for rocks of the Santa Rosa superunit in the Nepeña and Huaura valleys

look very different from those of the Lima segment.

The reasons for the more pervasive hydrothermal activity and mineralization in the Arequipa segment are not yet clear, but it is certainly true that the environments were different. Thus, the plutons in the north were emplaced and cooled in a thick sequence of emerging, marine, burially-metamorphosed volcanics (Aguirre *et al.*, 1978; Offler *et al.*, in preparation), while the country rocks in the south are Jurassic–Cretaceous volcanics and sediments which were probably above sea level during intrusion of the magmas. Work in progress will help to determine whether these differences in gross structural environment are sufficient to explain the variations in water–rock interaction.

Apart from the superficial differences caused by water–rock interaction, there are more fundamental differences between the two segments. Thus, the superunits of the Arequipa segment are more potassic than those of the Lima segment (Figures 10 and 11). They crystallized initially with more basic compositions (Figure 9), the average volumetrically weighted composition is more

Table 2 **Differences between the Lima and Arequipa segments of the Coastal Batholith of Peru**

	Lima	Arequipa
Ring complexes	Four	None
Country rocks	Volcanics (Albian)	Volcanics and sediments (Albian, Neocomian, U. Jurassic)
Superunits	8–9	5
Age of granitoids	> 95–61 Ma	> 97–81 Ma
Granites	Common	Rare
Aspect of country rocks	Usually unaltered	Reddened and commonly altered to epidote, chlorite and sericite
K–Ar dates	Some resetting	Comprehensive resetting
Mineralization	Slight	Extensive
Major element chemistry	K_2O content SiO_2 content	$< K_2O$ content $> SiO_2$ content

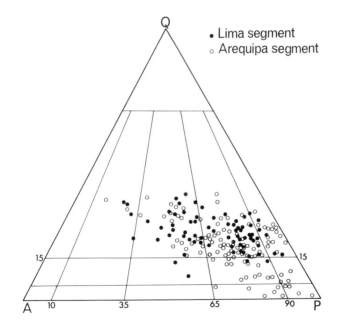

Figure 9 QAP diagram of the modes of rocks of the Lima and Arequipa segments, illustrating the presence of more basic rocks in the latter

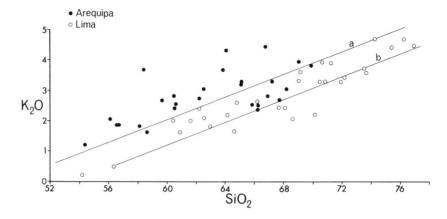

Figure 10 Plot of K_2O against SiO_2 for rocks of the Lima and Arequipa segments; a and b are lines limiting Arequipa and Lima data

basic and there is an absence of 'granites' (Figure 3). Indeed, the less extended differentiation and consequent lack of granitic end-members is an important feature distinguishing the Arequipa segment. It may be due to different initial water contents of the magmas or the quenching of the magmas of the Arequipa segment as they hit meteoric water before they had a chance to differentiate to granitic end-members and form ring complex plutons. However, there are no major consistent differences in trace elements, including the REE, between the plutonics of the two segments. Typically, the light REE are fractionated and the heavy REE are unfractionated in both segments. The basic units, which approach the compositions of the original liquids, have similar REE patterns and the differences which occur within the superunits can be explained by low-pressure fractionation involving mainly plagioclase, clinopyroxene and hornblende (Figure 4).

These features of the two segments, together with differences in mineralization and the crude coincidence of the structural and plutonic segmentation (Pitcher, 1978), may relate to the configuration of the descending oceanic plate beneath the Andes, which may, as now, have been different in each segment.

THE GABBRO ENIGMA

The relation of the gabbros to the granitoid suite and to the Casma volcanics is not clear. Certainly there is some overlap in time with the volcanics but the final emplacement of the granitoids is somewhat younger (Figure 2). Considering the Casma/gabbro relationship, both rock groups crop out along the length of the batholith, as far as we know without regard to segmental differences. But there is no simple plutonic/volcanic equivalence as the gabbro complexes do not contain equivalents of the more fractionated Casma volcanics. The more differentiated members of the gabbro suite are similar to the Casma basalts and one of us (W.J. McC.), considers the Casma basalts to represent the liquid fraction after crystallization of the gabbros. Unfortunately, the compatible element contents (Ni, Cr, Co

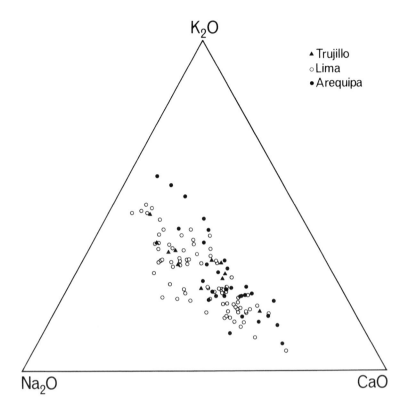

Figure 11 Na_2O-K_2O-CaO plot of rocks from three segments of the Coastal Batholith

and V) of the gabbro suite are indistinguishable, on our limited sampling, from the Casma basalts. Hence, we believe the Casma has no direct connection with the gabbros and is likely to have come from 'fissure eruptions, in a continental environment, along the axis and flanks of the eugeosyncline' (Webb, 1976).

A direct relationship between the gabbros and the granitoids is also unlikely as their ages are different. However, if the gabbros crystallized at depths of up to 35 km along the whole length of the batholith, as suggested by Regan (1976), there is some possibility that the 'granodioritic' superunits are the liquids from similar cumulate bodies of the requisite age at or near the base of the crust. These bodies and some of the differentiates could then account for some of the crustal thickening. However, these speculations need testing before the gabbros can be properly fitted into the batholith context.

CALIPUY VOLCANICS – RING COMPLEXES

The model whereby batholiths 'crystallize under a cover of their own volcanic ejecta' (Hamilton and Myers, 1967) is an attractive one and there is often no doubt about the temporal and compositional association. In Peru the beautiful ring dykes, gas action, tuffisite formation and the climax of activity in the ring complexes at about 62 Ma ago, led many workers (Knox, 1974; Bussell, 1975; Wilson, 1975; Bussell *et al.*, 1976; Pitcher, 1978) to postulate that the Calipuy volcanics originated from the ring complexes. The reasons we now think this is not so are outlined below, and it can be seen that the geological evidence is *paramount.* Thus, there are no ring complexes in the south, although the Calipuy volcanics extend well south of the Lima segment. The geochronologic work of Wilson (1975) and others indicates that the Calipuy is younger than the ring complexes, although it is true to say that the 'minimum' age of the oldest Calipuy (52.5 Ma) is near to that of the ring complexes (Figure 2). However, by far the greater part of the Calipuy is far too young. There is no evidence that the ring complexes cut the erosion surface on which the Calipuy rocks lie, and large-scale studies of the outcrop of the Calipuy indicate that its upper part at least may be related to the line of Oligocene–Miocene stocks, which crop out to the east of the batholith. The upper formation has an ignimbritic character, which is compatible with extrusion from these stocks (Webb, 1976).

The well-developed granophyric textures and compositions of the ring dykes indicate massive volatile release at the level of intrusion. Fine-grained equigranular equilibrium matrix textures in porphyritic rocks plus vesiculation, tuffisitization, etc., all confirm massive venting of gas as the advancing ring structures above the magma presumably neared or even broke through the surface. This separation of the vapour phase as the magma reached saturation very near to the surface (Atherton and Brenchley, 1972) effectively quenched the granitic magma at the present level, and explosive eruption of the ring dyke magma clearly did not occur. Indeed, in the case of Puscao, a structurally low component of the Huaura complex, there is an aplogranite facies which clearly formed at a depth at which degassing and magma venting to the surface could not occur. The magmas effectively 'died' at their present level. However, the evidence of quenching in the ring dykes does not preclude hydrothermal activity at the surface.

Other arguments making reconciliation of the volcanic–plutonic continuity difficult include the basic nature of the lower Calipuy, which is at variance with the clear acid character of the ring complex rocks. Indeed, the ring complexes, as one might expect from their eutectoid type compositions, are more fractionated than any of the acid Calipuy. Finally an interesting point, common to other batholiths, is the lack of contemporaneous basic plutonics (e.g. Carmichael *et al.,* 1974), which would be expected if batholiths and associated volcanics have a common source, or are different physical aspects of the same magma differentiation process.

Origin of the batholith and associated rocks

CASMA VOLCANICS

In the Lima segment, the Casma volcanics can be crudely divided into a western part, mainly made up of basalts and andesites and an eastern part, the Churin group, which is mainly dacites and rhyolites (Table 1 and Webb, 1976). Although at first sight this looks like a classic K-h relationship, there is no correlation between stratigraphic position, chemistry and age; thus, the dacites and rhyolites of the Churin group are older than many of the western basalts, a situation which clearly suggests a polymagmatic origin for these rocks, with no consistent increase in acidity with age.

The western Casma volcanics are low-K tholeiites with a primitive trace element chemistry, similar to island arc tholeiites (Jakeš and White, 1972), but erupted on submerged continental crust. They have low K_2O (0.21–0.65%), low K_2O/Na_2O ratios (0.06–0.34), low incompatible trace element and REE concentrations and little REE fractionation (Ce_N/Yb_N = 2–3; Figure 4). Ti–Zr, and Ti–Zr–Sr plots (Pearce and Cann, 1973) also indicate low-K tholeiite characteristics. These basalts are high-alumina in type and their trace element characteristics are very similar to rocks described, albeit of different age and structural environment, from the Western Stratovolcano Belt (37°–41°30' S), of Chile (Lopez-Escobar et al., 1977) and our preliminary model for their genesis would be very similar to theirs. That is, partial melting of spinel peridotite followed by fractional crystallization of olivine and clinopyroxene.

The eastern Casma volcanics are more highly fractionated, though they are the *same* age as the western group. They are more potassic and siliceous, with Rb/Sr ratios between 1.7 and 2.4. The lavas have high total REE levels, which are highly fractionated ($Ce_N/Yb_N \sim 6$) and have large negative Eu anomalies ($Eu_N/Eu_N^* \sim 0.5$) – see Figure 4.

The eastern Casma does not appear to be related to the western, and cannot be stratigraphically related by differentiation to the basalts. From our limited data and comparison with Chilean data (Klerkx et al., 1977), we have previously suggested a coherent crustal source for these rocks. This would be compatible with the spatial isolation of the acid volcanics. However, the REE data indicate that the acid rocks of both the Casma and the Calipuy had similar origins. The implications of this are considered in the following section.

CALIPUY VOLCANICS

The Calipuy is a thick variable succession, with a lower group of basic to intermediate lavas and tuffs and an upper group of acid pyroclastics (Table 1 and Webb, 1976). The basalts and andesites are true calc-alkaline rocks (Jakeš and White, 1972) with a high alumina composition. They show characteristic smooth differentiation trends on conventional variation diagrams, but are considered to be polymagmatic. The basalts, in particular, show differences from section to section, although alteration of the Calipuy is extensive and accounts for some marked variation in the more mobile elements, such as Rb. The dacites, particularly, possess an irregular chemistry due to extensive alteration.

Generally, the basic rocks show higher K and incompatible element contents than the Casma equivalents. They also have higher total REE, with enrichment in light REE ($Ce_N/Yb_N \sim 6$), and limited fractionation of the heavy REE ($Tb_N/Yb_N \sim 1$). The dacites are highly fractionated and have REE characteristics very similar to the Casma dacites (Figure 4).

The origin of the Calipuy basalts is, we believe, similar to that proposed for the Casma basalts but involving either smaller degrees of partial melting, a more

LIL-enriched mantle source or a different source mineralogy in order to account for the less primitive character of these rocks. The origin of the dacites is difficult to model due to alteration, but the high Zr content and general geochemical characteristics led one of us (McCourt, 1978) to suggest a crustal origin (cf. El Hinnawi *et al.*, 1969).

If we consider the spatial outcrops of the two groups of volcanics in relation to their tectonic setting, we find that the basic Casma occurs nearest to the trench while the Calipuy crop out further inland. The major and trace element contents of both sets of basic volcanics indicate a K-h relationship which we believe to be the result of decreasing degrees of partial melting with distance from the trench, a feature recognized in Chile (Lopez-Escobar and Frey, 1976).

The Calipuy dacites may not be of crustal origin as suggested above, because the REE data indicate they could be derived from a basic magma by fractional crystallization involving plagioclase. They also have very similar REE patterns to the Casma dacites, and hence both sets may be mantle-derived fractionates. The Calipuy basalts and dacites, plus the Casma dacites were all extruded along the same Andean line to the east of the basic Casma, and the REE evidence suggests the original magmas could have been the result of similar degrees of partial melting.

We believe the origins of the Casma and Calipuy are similar to those put forward by various workers in Andean Mesozoic and Cenozoic volcanism, e.g. James *et al.* (1976), Francis *et al.* (1977), McNutt *et al.* (1975) and Noble *et al.* (1974). The models relate to a subduction zone descending beneath the Andes, a situation we believe extended back to the time of formation of the two groups of volcanics under discussion here. However, the volcanics sampled are mainly basaltic rather than andesitic, a feature which may be related to a different geometry of the ocean crust/wedge in the Cenozoic. Our reasoning also leads to the conclusion that the dominant mechanism of extrusion of both sets of volcanics was via fissures in a continental slab parallel to the batholith axis (cf. Webb, 1976).

GABBROS

The average major element composition of the gabbros (Table 1) closely approximates to that of olivine tholeiite (Green *et al.*, 1967). The trace element data show low Rb (6 ppm), low Rb/Sr (0.01) and high Sr (*ca* 430 ppm). Total REE contents are low and there is little relative REE fractionation ($Ce_N/Yb_N = 1.2–3.0$). Some rocks show positive Eu anomalies ($Eu_N/Eu_N^* \sim 1.5$), suggesting they are in part plagioclase cumulates (Figure 4). These data are compatible with the work of Regan (1976) and Mullen and Bussell (1977), who proposed olivine tholeiite as parent magma for the Peru gabbros.

The general model for olivine tholeiite production, which is probably most applicable to the Peru gabbros, is partial melting in a mantle wedge, with the residual mineralogy not including garnet. The process may be initiated by the rise of volatiles from the descending plate and is followed by fractionation of olivine and clinopyroxene to explain the low transition metal contents (Table 1).

GRANODIORITIC AND GRANITIC PLUTONS

The problem of the origin of the granitoid plutons is the main theme of this paper and in some ways the most difficult to answer. The model depends crucially on certain pivotal interpretations of the data. Features which are clear and must have an important bearing on petrogenesis are:

1. The variations seen on the AFM diagrams (Figure 5) can generally be explained by high-level crustal differentiation involving hornblende, plagioclase, etc.
2. The individual superunit magmas of granodioritic, or more basic composition were formed *separately* as individual fusion/crystallization events.
3. Except for the gabbros, whose direct connection to the other plutons is unclear, there are no basic plutons.
4. A crucial feature is the primary basic—acid sequence (Pitcher, 1978) upon which the secondary individual differentiation paths are imposed. Does it exist?, or rephrased, are the late acid rocks of the ring dykes and associated late-stage granites individual melts as in (2), or high-level differentiates of a magma, the more basic part of which is not exposed?

If a primary basic-acid sequence is accepted, then a model of melting at higher and higher levels, starting in the mantle and ending in the upper crust, may follow (Brown, 1977; Pitcher, 1978). However, we think that the geological reasoning for this is at least questionable. Thus, superunits in the Lima segment, making up 92% of the batholith (omitting the gabbros), show little if any increase in acidity with decreasing age (Figure 3). The 'trend' is dependent on the ring complex superunits and the late K-feldspar phenocrystic plutons seen in the Lima segment *only* being 'primary' melts. The superunits making up the ring complexes (La Mina, Puscao and San Jerónimo) show a similar, though sometimes more acid, high-level differentiation as the earlier superunits and are, we believe, the more acid differentiates of similar magmas at depth. Indeed, despite the firmly held view of a primary basic to acid sequence, Pitcher (1978) indicates that 'it is very easy to show in the field . . . that the multipulse plutons evolved by drawing up magma from *underlying differentiating magma chambers* . . . the magma being displaced upwards around descending central blocks during the operation of cauldron subsidence'. As the acid magmas are clearly the most mobile phases during the differentiation, often breaking through the earlier members of a superunit, it seems logical to conclude that the ring members are end-members of similar granodioritic superunits equivalent to the Santa Rosa. Indeed, an underlying magma chamber must be axiomatic in a ring dyke model. Furthermore, the San Jerónimo granophyric superunit is unlikely to be the direct result of a fusion event in the lower crust, for the melting composition would be different at these depths.

The status of the late-stage granites is problematic, although Pativilca is apparently of a very different age to the ring complexes. Sayán and Cañas are less easy to explain for Sayán, like Pativilca, is a K-feldspar phenocrystic granite and neither it, nor Cañas or equivalents, are present in the ring complexes other than the Huaura. We conclude that they are exotic but structurally associated with the Huaura complex. Furthermore, as equivalent ring dyke acid rocks are not seen in the Arequipa segment, the acidity seems to be an artefact of ring complex formation and hence is a high-level phenomenon and not part of a *primary* sequence.

An important line of evidence with regard to the genesis of the plutonics is provided by the $^{87}Sr/^{86}Sr$ initial ratios. Values so far obtained are as follows: Santa Rosa 0.70411 ± 0.0001 (77 Ma); Puscao 0.70411 ± 0.00008 (65 Ma); and Sayán 0.70400 ± 0.00024 (64 Ma). These initial ratios clearly imply essentially mantle-derived material and, even more importantly, show no extra component of crustal material with decreasing age. From this evidence, they could well have all formed by high-level differentiation of mainly mantle material, as discussed above. If so, the granites in such environments are products of high-level differentiation, and the batholith cycle of Pitcher (1978) has nothing to do with 'the duration of a natural rhythm of crustal reworking' (Pitcher, p. 176) or increasing crustal involvement. Furthermore, the granitoid sequence shows a typical 'I' type character throughout and there is no tendency to 'S' type characteristics with decreasing age. Part of the attraction of the crustal involvement relates to the clear geological fact that the crust under the Andes is massively thickened to a maximum of 70 km, which in the extensional structural regime of Peru must be due to underplating (Brown, 1977; Pitcher, 1978) of mantle-derived material (Brown, 1977), either as a single or two or more stage process with little or no crustal involvement. The gabbros are surely an example of this accumulation.

Nevertheless, although the evidence suggests to us that the granitoids are in the main mantle-derived, the data do not allow us to discriminate completely against a low Rb crustal source, such as the Precambrian Mollendo gneisses in the Arequipa segment (which have a strontium initial ratio of 0.704 ± 0.002; Shackleton et al., 1979). Nor can we discount entirely the possibility of a crustal input which may have diluted a primitive mantle melt. However, this latter seems unlikely as the strontium initial ratios show little variation, so that the input would have to have been constant in each magma batch. More detailed work on the 'primary basic–acid sequence' using neodymium isotope ratios, etc., may help to elucidate the problem (see Hawkesworth, this volume).

In detail there are also problems with regard to the apparently more liquid character of the acid rocks, which carry less An-rich feldspar cores and basic clots than the more basic units, a feature which might indicate a significant difference between the earlier granodiorites and the later granites. However, we think this increasing mobility and separation is a function of fractional crystallization concentrating H_2O in the final liquid. The volume of tonalite required at depth may also be a problem, although it seems to us that batholiths such as the Coastal Batholith must have considerable extension at depth as we see no continuous change along the length of the batholith, which might relate to depth of emplacement. Finally, the problem of providing magmas of similar composition over a long time period from the mantle is an old one and still defies a complete answer as shown from the studies of andesite genesis.

At the moment, models of batholith genesis are numerous (cf. andesitic volcanics) and involve fractional crystallization, partial melting, vapour and/or liquid transport, magma mixing, etc. in a multistage process. Rarely can a semi-quantitative model be put forward even with good knowledge of partition coefficients, liquid–solid equilibria, fraction solidified, etc. As Gill (1978, p. 721) says, the models are 'non-unique, ad hoc, and await criteria for their selection' We believe that future models must be constrained by better field and petrographic evidence.

PETROGENETIC RELATIONSHIPS OF VOLCANIC AND INTRUSIVE ROCKS OF THE ANDES

R.S. THORPE and P.W. FRANCIS
Department of Earth Sciences, The Open University,
Walton Hall, Milton Keynes MK7 6AA, UK

Introduction

There are three distinct zones of recent volcanic activity in the Andean Cordillera: a northern zone in South Colombia and Ecuador, a central zone in South Peru and North Chile, and a southern zone in South Chile (Figure 1). Cenozoic intrusive rocks occur between these areas, and on the coastal margin of the active zones. It is therefore natural to suppose that the igneous processes that take place along the Andean plate margin have simultaneous volcanic and plutonic components and, consequently, that the two kinds of magmatic activity are interrelated. This supposition is expressed in countless diagrams and cross-sections purporting to illustrate destructive plate margin processes.

The spatial relationship between volcanic and plutonic suites and subduction zone processes in the Andes is self-evident. In addition, there are broad geochemical similarities expressed in major element composition in AFM and $K_2O:SiO_2$ plots between volcanic and intrusive rocks (cf. Figures 1 (b) and (c) and Pitcher, 1978, Figure 7 (a)). These geochemical similarities support the intuitive assumption that volcanic activity is the surface expression of plutonic processes.

When the evidence is examined more closely, however, a much less simple picture emerges. Thus, while Bussell et al. (1976) have confidently attributed the ring complexes of the Peruvian Coastal Batholith to a high-level crustal environment (Figure 2), with a direct link between the magmas of the constituent plutons and caldera-centred volcanicity, there are no *unambiguous* links with volcanic rocks. Furthermore, while Pitcher (1978) has argued that some of the pyroclastic flows making up part of the volcanic Calipuy Group (which forms the cover to the batholith) were erupted from the centred complexes, and Myers (1975) has suggested that the 'plutons rose into their own ejecta', Atherton et al. (this volume) have used geological and geochemical evidence to show that the volcanic rocks of the Calipuy Group are *not* consanguineous with the plutonic suites.

Geological evidence for links between volcanic and plutonic rock suites

We begin by examining the Peruvian Coastal Batholith, which is widely supposed to serve as a model for what underlies the present active volcanic belt in the Andes (Hamilton, 1969).

The arguments for the links between volcanic and plutonic suites in the Peru

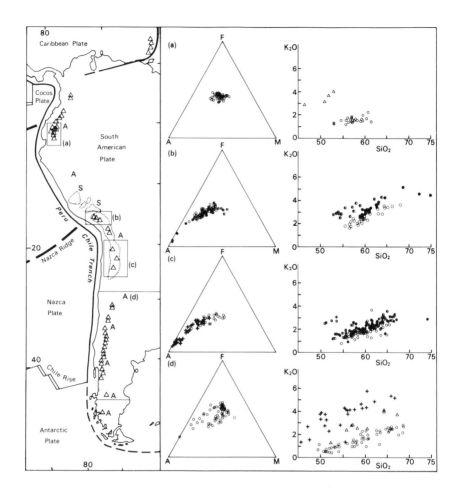

Figure 1 The distribution and compositional characteristics of active volcanoes, and their relationships to plate tectonics in the South American Andes. The map shows active volcanoes (open triangles) in relation to plate tectonic features (active volcanoes are from Macdonald (1972) with the addition of Volcan Hudson in South Chile). Solid lines are destructive plate boundaries; thin paired lines are constructive boundaries; thick broken lines are oceanic ridges and/or rises in the Nazca plate and the broken line (south of $42°S$) is the boundary of the Antarctic and South American plates. The dotted line outlines the ignimbrite province of the Central Andes; S = shoshonitic and A = alkaline volcanic rocks. The AFM diagrams and plots of K_2O against SiO_2 ((a)–(d)) are for rocks from the areas indicated on the map and are as follows:

(a) Ecuador, calc-alkaline lavas (open circles; Francis and Thorpe, unpublished; Pichler *et al.*, 1976) and alkaline lavas from Sumaco (open triangles; Colony and Sinclair, 1928).

(b) South Peru, calc-alkaline and shoshonitic lavas. Open circles = A_1 series; solid circles = A_2 series; half-filled circles = B series (cf. Lefévre, 1973, Figure 4).

(c) North Chile, calc-alkaline and shoshonitic lavas. In the AFM diagram, open circles = lavas (Francis *et al.*, 1974); crosses = ignimbrites (Francis *et al.*, 1974; Thorpe *et al.*, 1979). In the K_2O vs. SiO_2 plot (cf. Roobol *et al.*, 1976, Figure

(contd opposite)

WSW   ENE

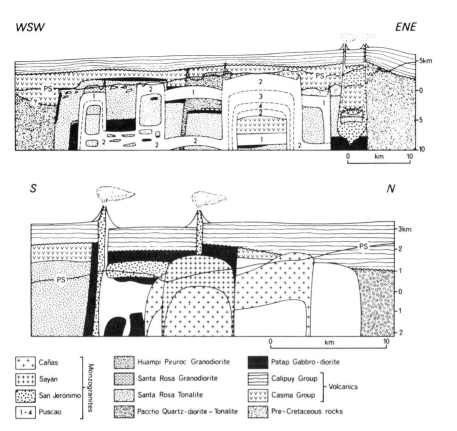

+ + + Cañas		Huampi Piruroc Granodiorite
+ + + + Sayán	Monzogranites	Santa Rosa Granodiorite
San Jerónimo		Santa Rosa Tonalite
1 - 4 Puscao		Paccho Quartz-diorite – Tonalite

Patap Gabbro-diorite	
Calipuy Group	Volcanics
Casma Group	
Pre-Cretaceous rocks	

Figure 2   Schematic sections across the Coastal Batholith of Peru showing postulated relationships between intrusive and extrusive rocks. (Top) Section along the northern flank
of the Fortaleza Valley (after Myers, 1975). (Bottom) Section along the southern
flank of the Huaura valley (Bussell *et al.*, 1976). PS = present erosion surface

batholith have rested mainly on the strong similarities between the ring complexes
in the batholith and those elsewhere in the world where cauldron subsidence and
eruption of voluminous pyroclastic flows have taken place. Pitcher (1978, p. 174)
states that: 'There is no doubt that these complexes are volcano-plutonic forma-
tions . . . representing a direct connection between the intrusives of the batholith

Figure 1 (contd)

2), open circles = Western Chain lavas; filled circles = Eastern Chain lavas; right-
filled circles = Argentinian lavas; left-filled circles = Bolivian lavas.
(d) South Chile, calc-alkaline and alkaline lavas. The AFM diagram is for calc-
alkaline lavas from Moreno (1976), and Lopez-Escobar *et al.* (1976, 1977).
In the K_2O vs. SiO_2 plot, open circles = calc-alkaline lavas (Deruelle and
Deruelle, 1974; Lopez-Escobar *et al.*, 1976, 1977); open triangles = alkaline
lavas from Pino Hachado (Vergara, 1972; Lopez-Escobar *et al.*, 1976) and
Volcan Hudson (Ponce, 1976); crosses = alkaline lavas from Pocho and Payun
Matru in North-west Argentina (Vergara, 1972)

and the volcanics of the country rocks and cover' and, in addition, such a 'direct pluton-volcanic connection has persisted throughout the entire history of the granitoids' (cf. Figure 2).

The most extensive and comprehensively studied cauldron structures are located in the south-western USA, where pyroclastic flows with volumes totalling tens of thousands of cubic kilometres have been mapped, and many separate cauldrons located, each of which is supposed to be located over a pluton. No direct links with plutonic rocks have been observed, however, notwithstanding the fact that the cauldrons are relatively old (*ca* 20–30 million years) and deeply eroded (Elston *et al.*, 1975a). In describing the Mogollon Plateau volcanic ring complex as 'the surface expression of a major batholith', Elston was forced to conclude that 'no single point proves that the pluton exists, but the wide spectrum of evidence strengthens its plausibility' (Elston *et al.*, 1975b).

Thus, something of a paradox emerges. Spectacular examples of plutonic ring complexes are found which appear to lack obvious volcanic counterparts, while major volcanic cauldron structures cannot be positively correlated with a plutonic suite.

In the active volcanic region of the Central Andes, two main components can be recognized, each of which may have a distinct plutonic counterpart. By far the most abundant are the andesite composite volcanoes of which there are nearly 200 between latitudes 19° 30' and 23° S (Baker, 1974). Typically, these are cones reaching up to 3000 m above local base level, and each has a volume approaching 100 km³ (Baker and Francis, 1978). Each cone must clearly have a plutonic 'feeder', and may be located above its own 'magma chamber'. Such a magma chamber might correspond with the individual granite and tonalite plutons, each of which has a diameter of *ca* 10–20 km, which constitute the Peruvian batholith. The only volcanic expressions of plutonic activity that are clearly recognized by Pitcher (1978) are calderas and ring dykes associated with the centred complexes, but there is no evidence from the active volcanic belt that the andesite composite volcanoes are associated with either calderas or ring dykes.

Large caldera complexes or 'cauldrons' and ring structures are not abundant in the active volcanic province, but where they occur they are quite distinct from the andesite volcanoes. An example of a caldera with a resurgent centre, surrounded by voluminous pyroclastic flows, is the Cerro Galan caldera of North-west Argentina (Francis *et al.*, 1978). Other major calderas are Cerro Panizos on the Bolivia-Argentina frontier (Kussmaul *et al.*, 1977) and Cerro Purico on the Chile-Bolivia border (Francis *et al.*, unpublished). Several others are known in the eastern Cordillera of Bolivia (Francis *et al.*, 1978). Each of these structures is associated with many tens or hundreds of cubic kilometres of pyroclastic flows, and re-sembles so-called 'cauldron' structures, but is *not* linked with andesite composite volcanoes.

It is tempting to draw links between the large cauldrons and the ring structures in the Peru Batholith. There is a problem of scale, however. The Cerro Galan caldera, for example, is 40 km in diameter – nearly three times the size of the Peruvian ring complexes. A few small structures are known with diameters of between 5 and 10 km which may be the surface expression of ring dykes, but these are not associated with any significant volume of extrusive rocks (see, for example, the circular structure noted in Francis and Baker (1978)).

It is worth noting that clear examples of 'cauldron' structures associated with large volumes of pyroclastic flows are best developed on the eastern Cordillera of the Andes. In the western Cordillera, where most active andesite volcanism is concentrated, large volume pyroclastic flows are found, but there are few simple caldera structures. The source of many of the large pyroclastic flows is quite unknown, but there has long been a suspicion (see, for example, James (1971)) that the pyroclastic flows were erupted from long linear fissures. If this were the case, then presumably there should be linear – or at least elongate – plutonic equivalents. No such linear bodies have been described from the Peruvian batholith, where individual plutons all appear to be roughly equant.

One simple explanation for the apparent absence of links between volcanic and plutonic rock suites may lie in the quite different rates at which they accumulate. Estimates of the rate at which batholiths form are scarce, but Francis and Rundle (1976) showed that the Peruvian Coastal Batholith has accumulated at a rate broadly comparable to that at which volcanic rocks have been erupted over the last 20 million years. Whereas the batholith reveals the time-integrated accumulation of intrusive activity over 70 million years (Pitcher, 1978), the volcanic rocks exposed at the surface are mostly less than 20 million years old and, clearly, much has been lost by erosion (Baker and Francis, 1978). It is conceivable that in a further 10 million years no volcanic rocks at all will be preserved at the surface, particularly if the climate were to become less arid than it is today. Thus, assuming that plutonic rocks are being intruded beneath the present active volcanic belt, there is no certainty that any signs of a volcanic 'cover' will remain when the plutonic rocks are exposed.

Much of the preceding discussion has been based on the premise that the Peruvian batholith was formed by magmatic activity at a destructive plate margin at the same time as volcanic activity was proceeding at the surface and that, by implication, the present active volcanic belt of the Central Andes should be underlain by a Peruvian batholith-type counterpart. Is this premise necessary? Would it be possible for large volume intrusions to be emplaced without a volcanic expression and, conversely, would it be possible for large volumes of volcanic material to be erupted without an intrusive counterpart? We look first at the petrogenesis of Andean andesite, before returning to this problem.

Andean petrogenetic processes

The most voluminous and characteristic volcanic group of the Andes is the high Al-basalt—andesite—dacite group which is responsible for building the major composite volcanoes of the Cordillera. Several characteristics of the association indicate a predominantly mantle-derived origin. Firstly, the restriction of active volcanism to three areas with more steeply dipping Benioff Zones than volcanically-inactive areas clearly indicates that the petrogenesis of the volcanic rocks is linked with the presence of a well-developed mantle wedge of asthenospheric material (Baranzangi and Isacks, 1976, 1979). Secondly, Sr- and Nd-isotope data for Andean andesites are consistent with an origin from an oceanic crust—mantle system (Hawkesworth *et al.*, 1979a; Hawkesworth, this volume). Finally, it has been noted that, while substantial crustal thickening has taken place in the Central Andes, this has not resulted from regional compressive tectonics but must reflect addition of magmatic material from the mantle (James, 1971).

We therefore accept that the Andean andesite association results from partial melting or fractional crystallization of mantle-derived magmas. Possible sources are therefore subducted oceanic crust and the overlying mantle wedge. However, if the lower crust consists of young accreted magmatic material, a contribution from this source may be difficult to distinguish from subcrustal material. Within the plate tectonic framework of the Central Andes, magmatism would be initiated by partial melting or dehydration along the Benioff Zone. Partial melts of oceanic crust may have the broad characteristics of andesite but must either react with the overlying mantle or be modified by fractional crystallization to reach the surface as andesite (Stern, 1974; Fyfe and McBirney, 1975; Stern and Wyllie, 1978).

Although there is some disagreement about the depth of possible melting along the subduction zone, it seems likely that dehydration of the descending oceanic crust will take place at depths of 70–120 km and possibly deeper (Mysen, 1978; Anderson *et al.*, 1978). Such dehydration of oceanic crust would cause enrichment of the mantle wedge in a range of elements, including SiO_2 and large ion lithophile (LIL)-elements (Anderson *et al.*, 1978). In either case the enriched mantle would be the source for parental andesite magmas (cf. Ringwood, 1974; Thorpe *et al.*, 1976; Dostal *et al.*, 1977a, 1977b; Hawkesworth *et al.*, 1979a). In this context the petrogenesis of andesite might involve some or all of the following processes:

1. Partial melting of enriched mantle above the subduction zone.
2. Fractional crystallization of these magmas at mantle depths.
3. Contamination of these magmas with continental crust.
4. Fractional crystallization of olivine, pyroxene and ($<$ 35 km depth) plagioclase *within* the continental crust.

A range of chemical data can be used to infer details of (1) to (4) above.

For Central Andean andesites, rare earth element (REE) data are consistent with an origin by partial melting of 'enriched' garnet bearing peridotite (Thorpe *et al.*, 1976; Dostal *et al.*, 1977a, 1977b). Although such a source may be enriched in LIL elements, it would also have high Ni and Cr concentrations comparable to those estimated for mantle elsewhere. The parental magmas for andesites must therefore have had high Ni and Cr concentrations, and high pressure olivine ± pyroxene crystallization would be required to reduce concentrations of these elements to the low values characteristic of andesites.

For andesite in South Chile, Lopez-Escobar *et al.*, (1977) suggested an origin by 3% partial melting of garnetiferous peridotite (containing 6% garnet) followed by 20% fractional crystallization of olivine (80%) and clinopyroxene (20%), and similar models can be applied to andesite lavas in North Chile. Following crystallization of olivine, extensive crystallization of pyroxene + plagioclase is indicated by Rb, Sr and REE data (Thorpe *et al.*, 1976, 1979).

The model outlined above can be used to account for regional variation in Andean andesite composition. The K-h relationship might be explained by a decrease in degree of partial melting and/or an increase in degree of fractional crystallization on passing from west to east (Thorpe and Francis, 1979). The longitudinal variations in andesite composition can also be explained by the same processes. Basaltic andesites and andesites from Ecuador are more basic than those in North Chile (Pichler *et al.*, 1976; Francis *et al.*, 1977), but have similar REE abundances and patterns. In view of the similar depth to the Benioff Zone and the thinner continental crust, these andesites might have formed by similar processes to those of North Chile, but have experienced less fractional crystallization during ascent. The high-alumina basalts and basaltic andesites of South Chile probably formed at a shallower depth and as a result of a higher degree of partial melting or a lower degree of fractional crystallization (Lopez-Escobar *et al.*, 1977).

North Chilean andesites from the San Pedro—San Pablo volcano have higher and more variable $^{87}Sr/^{86}Sr$ ratios (0.706–0.707) than have basaltic andesites and high-alumina basalts from Ecuador and South Chile (*ca* 0.704, Francis *et al.*, 1977; Klerkx *et al.*, 1977). These high $^{87}Sr/^{86}Sr$ ratios are generally attributed to a small component of crustal contamination (Francis *et al.*, 1977; Briqueu and Lancelot, 1979). The same lavas also have lower $^{143}Nd/^{144}Nd$ ratios (0.5124–0.5126) than Ecuadorian basaltic andesites (0.5129–0.5130; Hawkesworth *et al.*, 1979a) and, if contamination has occurred, then both Sr and Nd must have been affected. This problem is evaluated in more detail below.

The role of continental crust

The rise of basaltic or andesitic magma (density = 2700–2800 kg m^{-3}) from the upper mantle (density = 3300 kg m^{-3}) into thickened continental crust (density = 2700–3000 kg m^{-3}) might slow the progress of mantle-derived magmas towards the surface, causing such magmas to underplate or intrude the continental crust. Under such circumstances, calc-alkaline magmas might experience more fractional crystallization or contamination than magmas emplaced into thin crust. Sr- and Nd-isotope data for andesites from Ecuador (crustal thickness = 40–50 km) and North Chile (crustal thickness = *ca* 70 km) can be used to evaluate these possibilities (cf. Francis *et al.*, 1977).

A plot of $^{143}Nd/^{144}Nd$ vs. $^{87}Sr/^{86}Sr$ for andesites from Ecuador and North Chile and for a range of continental rocks is shown in Figure 3. The figure shows the range of Nd—Sr isotopic variation in the mantle and in continental rocks from North America (McCulloch and Wasserburg, 1978) and Scotland (Carter *et al.*, 1978; Hamilton *et al.*, 1979). Also shown are lines indicating Nd- and Sr-isotopic variations in mixtures of mantle-derived and crustal rocks (cf. DePaolo and Wasserburg, 1979). Such mixing lines generally have gentle curvature since common crustal contaminants have higher Nd/Sr than mantle-derived magmas (cf. DePaolo and Wasserburg, 1979).

To evaluate possible effects of crustal contamination of Andean andesite

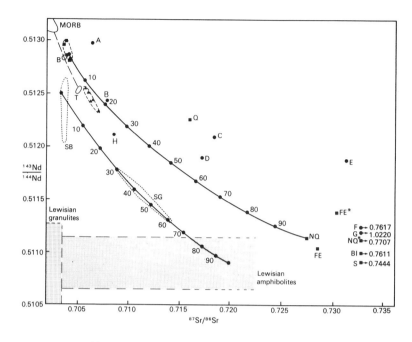

Figure 3 Plot of $^{143}Nd/^{144}Nd$ vs. $^{87}Sr/^{86}Sr$ for Andean calc-alkaline lavas and various continental rocks. The 'mantle array' is indicated by the dashed line joining mid-ocean ridge basalt (MORB), Bouvet (B) and Tristan (T) (O'Nions *et al.*, 1977). Unlabelled filled squares are Ecuadorian andesites and filled triangles are North Chilean andesites (Hawkesworth *et al.*, 1979a). The field labelled SB = Skye basalts and SG = Skye granites (Carter *et al.*, 1978). The coarse stippled field is granulite-facies Lewisian gneiss and the fine-stippled area is amphibolite-facies Lewisian gneiss (Carter *et al.*, 1978; Hamilton *et al.*, 1979, with additional Sr-isotope data). Filled squares with letters are averages for the Canadian Shield composites (McCulloch and Wasserburg, 1978) with NQ = North Quebec; NQ* = New Quebec; FE = Fort Enterprise Granite; FE* = Fort Enterprise Gneiss; BI = Baffin Island; S = Saskatchewan; Q = Quebec. Filled circles are sedimentary rocks with A = Baja Shale; B = Deep-sea Red Clay; C = Nanking Loess; D = Iowa Loess; E = North American shales; F = Birch Creek schist; G = Figtree shale; H = San Gabriel Sand (McCulloch and Wasserburg, 1978, Table 3). The upper solid line shows the effect of contamination of a rock with the isotope characteristics of Ecuadorian andesite EF20 (Francis *et al.*, 1977; Hawkesworth *et al.*, 1979a) with the North Quebec quartzofeldspathic gneiss composite (Shaw *et al.*, 1976; McCulloch and Wasserburg, 1978). The lower solid line shows the effect of contamination of Skye basalt P3 (Carter *et al.*, 1978) with a Lewisian amphibolite with Sr = 500 ppm; $^{87}Sr/^{86}Sr$ = 0.72; Nd = 40 ppm; $^{143}Nd/^{144}Nd$ = 0.5109 (see text for discussion). All data plotted and used in calculations have been normalized to $^{87}Sr/^{86}Sr$, E−A = 0.70800 and $^{143}Nd/^{144}Nd$, BCR−1 = 0.51263

magmas, data are plotted from Skye where there is general agreement that some of the granitic rocks include a crustal component, although it is not clear how this was incorporated into the granite magmas (cf. Moorbath and Bell, 1965; Moorbath and Welke, 1968; Forester and Taylor, 1978; Thorpe *et al.*, 1977; Carter *et al.*,

1978; Moorbath and Thompson, in press). Carter *et al.* (1978) used combined Nd–Sr-isotope evidence to show that 'there is no doubt that some of the Tertiary volcanics of North-west Scotland have been markedly contaminated by Lewisian granulite facies basement', that 'between 5 and 50% of the Sr and Nd in the various lavas has been derived from the Lewisian basement' and that 'Sr and Nd were extracted preferentially from basement rocks with respect to SiO_2' (Carter *et al.*, 1978, p. 745). They also argue that 50–75% of the Nd and Sr in six Tertiary intrusive rocks from Skye (including a gabbro) is derived from amphibolite facies Lewisian. The fields for Skye basalts and granites are given in Figure 3, together with a mixing line between a Skye basalt sample and an amphibolite-facies gneiss with the characteristics defined in the figure caption. It is clear from this that the granites may have 30–60% crustal Nd and Sr (cf. Carter *et al.*, 1978). However, it is also evident from the basalt data that such contamination is selective and does not affect the major element characteristics of the magma (cf. Moorbath and Thompson, in press).

Figure 3 also shows a mixing line between a rock with the Nd- and Sr-isotope characteristics of an Ecuadorian andesite and a continental composite from the Canadian Shield (McCulloch and Wasserburg, 1978). Although the isotopic characteristics of the North Chilean andesites may be consistent with formation from a mantle system enriched in Rb/Sr and Nd/Sm relative to the bulk Earth (see Hawkesworth, this volume), the data are equally consistent with formation from magma formed from a more 'normal' mantle system (such as that parental to the Eduadorian andesites) followed by contamination with 10–20% crustal Nd and Sr. In view of the arguments concerning the Skye basalts, such contamination would probably involve much smaller proportions of crustal rock and we propose that the Nd- and Sr-isotope characteristics of the North Chilean andesites result from contamination of parental andesite magma, with a small but isotopically significant amount of ancient low $^{143}Nd/^{144}Nd$, high $^{87}Sr/^{86}Sr$ continental crustal material. Crust with such characteristics may underlie the North Chilean Andes (cf. Shackleton *et al.*, 1979). Since the amount of contamination proposed is rather small, probably below 5%, we propose that the more SiO_2-rich character of the North Chilean andesites and the greater relative abundance of acid volcanic rocks in the Central Andes is *largely* a result of greater fractional crystallization of andesite magmas rising through thick continental crust in this area (Thorpe *et al.*, 1979).

Petrogenesis of Central Andean ignimbrites

Voluminous ignimbrites of dacite-rhyolite composition form a prominent part of the volcanic landscape of the central volcanic zone (Figure 1). It is often claimed that crustal fusion is important in the petrogenesis of such magmas (e.g. Pichler and Zeil, 1972; Fernandez *et al.*, 1973). However, few petrological characteristics favour an origin by crustal fusion and the overlapping initial strontium-isotope ratios of andesites and ignimbrite magmas suggest that these rocks share a common source (Thorpe *et al.*, 1979). The higher Rb/Sr ratios and greater Eu anomalies in the ignimbrites are consistent with an origin by plagioclase-dominated fractional crystallization of mantle-derived andesite magma. The restriction of dacite-rhyolite ignimbrites to the Central Andes is therefore consistent with the occurrence of extensive fractional crystallization during the slow rise of andesite magma through thick continental crust.

Petrogenetic links between intrusive and extrusive rocks

In the introduction of this paper we commented on the difficulty of defining geological evidence for the relationship between intrusive and extrusive Andean igneous rocks. While it would be unreasonable to divorce plutonic from volcanic rocks completely, it may be that in a given province they are separated in *time*. It is possible that during magmatic activity at a persistently active destructive plate margin, large-scale plutonic activity predominates in the early stages and is followed by volcanic and plutonic activity, so that the two processes only take place concurrently during the later stages of the evolution of the plate margin. We now explain this possibility further, using the petrogenetic model outlined above.

A possible explanation for the separation in time may lie in the water content of the magmas. It is well-known that a major control on the depth of crystallization is the water content of a magma (Harris *et al.,* 1970), such that only relatively 'dry' magmas are able to reach shallow crustal levels and be erupted as volcanic rocks. Pitcher (1978) has suggested that the plutons of the Peruvian batholith rose to within 3–8 km of the surface and froze to a stop when they absorbed water from the zone of meteoric circulation. Alternatively, the batholithic magmas may simply be those calc-alkaline magmas with higher primary water contents than the erupted andesites.

Evidence concerning the water content of andesitic magmas is rather difficult to obtain. However, using petrological data, Sakuyama (1979) has suggested that the water contents of andesite magmas increase away from the trench. He suggests that water behaves as an incompatible element and that the parental magmas for the andesites are highly water-undersaturated. This is consistent with our petrological model in that pyroxene-plagioclase-bearing andesites formed by relatively large degrees of fractional crystallization are clearly not water-saturated. The parent magmas must, therefore, have been water-undersaturated (cf. Marsh, 1976) and this would facilitate their movement to high levels within the crust and their subsequent eruption.

However, a paradox emerges when the ignimbrites are considered, since these contain hornblende and biotite and were erupted explosively, thus suggesting a considerable water content. For similar rhyolitic ignimbrites from New Zealand, Rutherford and Heming (1978) suggest a pre-eruption water content of 4–6%. We therefore suggest that the parental andesite magmas for the ignimbrites were initially highly water-undersaturated, thus enabling the magmas to penetrate close to the surface, but that the high degree of fractional crystallization inferred for their origin was also responsible for increasing the water content of the magma and possibly contributing to their explosive eruption. This is consistent with the argument of Cann (1970) that the higher an acid magma rises in the crust, the deeper is its likely origin.

These processes for uprise and crystallization might explain a temporal separation of plutonic and volcanic activity while still accounting for the broad chemical similarity of plutonic and volcanic rock suites. The early batholithic magmas may initially have been more hydrous or have interacted with meteoric water. Such interaction is indicated by mineralization around high-level intrusive stocks (Pearce, 1976). This activity *might* not have been accompanied by volcanism. However, later magmas would be drier either as a result of depletion of this source area in water, by exhaustion of the meteoric water supply or by intrusion through earlier plutonic rocks, and would hence penetrate closer to the surface, being erupted as lavas or pyroclastic rocks if extensive near-surface crystallization and water concentration takes place.

Conclusions

Accepting that there is a geological link between volcanic and intrusive processes, but that the two may not always take place concurrently, and noting the overall geochemical similarity between them, we propose that the petrogenetic model for the andesitic rocks is probably also applicable to the origin of the tonalitic magmas of the batholith.

We conclude, therefore, that both intrusive and extrusive suites of calc-alkaline rocks are mantle-derived, but that there is a small — and sometimes overstated — contribution from the continental crust. Together, they form a significant contribution to continental growth in the Central Andes. The behaviour of the magmas once formed, whether they ultimately form extrusive or intrusive rocks, is largely controlled by their water content which varies and which behaves in ways not yet fully understood.

$^{143}Nd/^{144}Nd$, $^{87}Sr/^{86}Sr$ AND TRACE ELEMENT CHARACTERISTICS OF MAGMAS ALONG DESTRUCTIVE PLATE MARGINS

C.J. HAWKESWORTH
Department of Earth Sciences, The Open University,
Walton Hall, Milton Keynes MK7 6AA, UK

Introduction

The isotope compositions of Sr, Pb and, more recently, Nd have been widely used in the study of magmatic processes along destructive plate margins. Empirically they may help to resolve whether particular magmatic associations are related to one another, but their real potential lies in the detection of different components which may contribute to magmas in this tectonic environment.

In an island arc sequence material might be derived from basalt and sediment in the subducted lithosphere and from the overriding mantle wedge. Moreover, the chemistry of the latter will depend on its evolution before subduction took place. Along a continental margin the picture is further complicated both by the possibility of crustal contamination and by potential differences in the chemistry of the subducted sediment and the nature of the continental lithosphere.

Pb-isotopes are beyond the scope of this review and for an introduction to their application to island arc magmas the reader is referred to two recent case studies on the Marianas (Meijer, 1976) and the Aleutians (Kay et al., 1978). Many results on magmas from destructive plate margins fall on steep slopes on diagrams of $^{207}Pb/^{204}Pb$ versus $^{206}Pb/^{204}Pb$ compared to data from constructive margin and intraplate magmas. This is interpreted in terms of mixing between 'continental' Pb in the subducted sediments and 'mantle' Pb, probably in the overriding mantle wedge. However, because ocean sediments have an average of 25 ppm Pb (Chow and Patterson, 1962) and island arc basalts have 1–3 ppm Pb, the postulated contribution from ocean sediments is usually less than 5%.

$^{87}Sr/^{86}Sr$ ratios

The Sr-isotope composition of most island arc magmas tends to be more radiogenic than that of mid-ocean ridge basalts (MORB) but is similar to that of many ocean island magmas. Faure (1977), in a recent compilation, calculated the mean $^{87}Sr/^{86}Sr$ ratios of the available results from the different environments: ocean floor, 0.7028; ocean island, 0.7039; and island arc, 0.7044. However, we should consider the question: are the processes responsible for such $^{87}Sr/^{86}Sr$ ratios in destructive plate margin magmas really comparable to those operating in the intraplate or constructive plate margin environment?

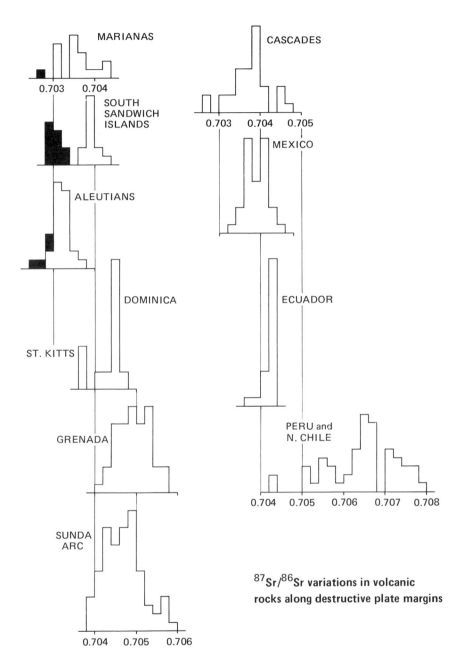

Figure 1 $^{87}Sr/^{86}Sr$ ratios from volcanic rocks erupted along destructive plate margins: those in black are from basalts in the 'back-arc' environment. Data from Church and Tilton (1973), DePaolo and Wasserburg (1977), Francis *et al.* (1977), Hawkesworth *et al.* (1977, 1979a, 1979c), Hawkesworth and Powell (1979), James *et al.* (1976), Kay *et al.* (1978), Meijer (1976), Moorbath *et al.* (1978), Noble *et al.* (1975), Whitford and Bloomfield (1975), and Whitford (1975)

Figure 1 summarizes $^{87}Sr/^{86}Sr$ ratios reported in some of the more detailed investigations of destructive plate margin magmas. The left-hand column portrays results from island arc magmas in oceanic areas which cannot therefore have been contaminated by continental crust *en route* to the surface. Two points should be emphasized:

1. Where analyses on magmas from behind the arc are also available, the island arc magmas have significantly *higher* $^{87}Sr/^{86}Sr$ ratios (Figure 1). This suggests that the subduction process has been responsible for an increase in $^{87}Sr/^{86}Sr$.

2. Island arc magmas can have $^{87}Sr/^{86}Sr$ ratios as low as 0.703 in the Marianas and the Aleutians, and up to almost 0.706 in Grenada and in the Sunda arc. Such variations might reflect the build-up of radiogenic Sr during subduction and/or differences in $^{87}Sr/^{86}Sr$ which were present before subduction took place.

The same two components – one from the subduction process and one from the overriding lithosphere – presumably contribute to the Sr composition along continental margins, and certainly their $^{87}Sr/^{86}Sr$ ratios are often very similar to those in oceanic island arc magmas (Figure 1). More controversial is whether the higher $^{87}Sr/^{86}Sr$ ratios in areas such as Peru and N Chile reflect higher Rb/Sr ratios during the pre-subduction evolution of the overriding lithosphere or contamination with continental crust.

Nd- and Sr-isotopes

MORB AND INTRAPLATE MAGMAS

The combination of Nd- and Sr-isotopes has proved to be a powerful tool in the study of magmatic rocks. Their decay schemes involve trace elements which have been well studied, and whose chemical characteristics are sufficiently diverse to suggest that their relative behaviour in a melt may be different from that in an H_2O-rich fluid. Results are generally presented on a diagram of $^{143}Nd/^{144}Nd$ vs. $^{87}Sr/^{86}Sr$, and the majority of recent mantle-derived rocks from mid-ocean ridges and both continental and oceanic intraplate environments plot on a single broad negative trend (Group I, in Figure 2).

DePaolo and Wasserburg (1976) and O'Nions *et al.* (1977) used this broad trend to estimate the present-day $^{87}Sr/^{86}Sr$ ratio of the model bulk Earth. If the Earth had chondritic Sm/Nd and initial $^{143}Nd/^{144}Nd$ ratios, then its present-day $^{143}Nd/^{144}Nd$ ratio is 0.51262. Assuming that Rb/Sr and Sm/Nd have behaved coherently throughout Earth history, and that the present-day isotope composition of the bulk Earth therefore lies on the main correlation between $^{143}Nd/^{144}Nd$ and $^{87}Sr/^{86}Sr$, its present-day $^{87}Sr/^{86}Sr$ ratio is 0.7045–0.7050 (Figure 2). Thus, most recent mantle-derived volcanic rocks have *higher* $^{143}Nd/^{144}Nd$ and *lower* $^{87}Sr/^{86}Sr$ ratios than the present-day composition of the bulk Earth. This implies that their source regions in the upper mantle have

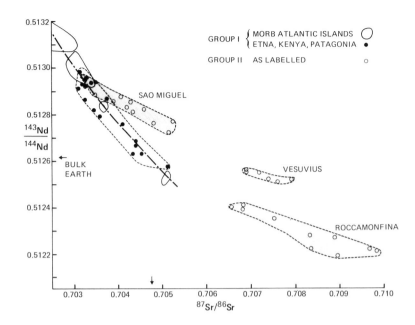

Figure 2 ^{143}Nd/^{144}Nd versus ^{87}Sr/^{86}Sr for recent basalts from mid-ocean ridges and continental and oceanic intraplate environments, after Hawkesworth *et al.* (1979b). Data also from Carter *et al.* (1978), Hawkesworth and Vollmer (1979), Hawkesworth *et al.* (1979a), Norry *et al.* (1979), and O'Nions *et al.* (1977). (From *Nature*, 1979, vol. 280, p. 29, Figure 3; by courtesy of Macmillan, London)

been relatively depleted in Rb and in light rare earth elements (LREE) for much of this history. However, many volcanic rocks from both continental and oceanic areas are characterized by high concentrations of large ion lithophile (LIL)-elements and they have clearly not been derived from a LIL-element depleted source. It is therefore necessary, when discussing the trace element characteristics of a portion of the upper mantle, to make a clear distinction between those deduced from the trace element chemistry of igneous rocks and those inferred from isotope composition.

In a review of ^{143}Nd/^{144}Nd and ^{87}Sr/^{86}Sr results on recent volcanic rocks, Hawkesworth *et al.* (1979b) tentatively recognized two groups (Figure 2). Group I is the main trend of mantle-derived volcanic rocks and most of them have isotope compositions indicative of derivation from source regions which were depleted in LIL-elements relative to the bulk Earth for a considerable period of time. Thus, the source of LIL-element-enriched magmas in Group I must have been only recently enriched in LIL-elements, and those elements probably migrated from source regions which had Nd- and Sr-isotopes that also plotted on the trend of the Group I samples.

Group II rocks, by contrast, have shallower slopes and plot to the right of the main correlation (Group I) on the graph of ^{143}Nd/^{144}Nd vs. ^{87}Sr/^{86}Sr. They have so far been recognized in only two areas (Figure 2) and clearly represent a minute fraction of recent volcanic rocks. However, their interest lies in the fact that they involve a different component to those in Group I. Hawkesworth *et al.* (1979b) have argued that LIL-element enrichment again took place comparatively recently but that in this case the elements were derived from material characterized by high $^{87}Sr/^{86}Sr$ and low $^{143}Nd/^{144}Nd$ ratios. Since these are often taken to be crustal characteristics, one possible explanation is that it might reflect subducted crustal material which had retained its identity in the upper mantle. Alternatively, Sr- and Nd-isotope results on xenoliths of metasomatized mantle from kimberlites (Erlank and Shimizu, 1977; Hawkesworth and Erlank, unpublished) demonstrate that mantle metasomatism can result in LIL-element enrichment which, in time, will generate the relatively high $^{87}Sr/^{86}Sr$ and low $^{143}Nd/^{144}Nd$ ratios apparently characteristic of the Group II end-members.

In summary, the available Nd- and Sr-isotope results on mid-ocean ridge basalts and intraplate magmas suggest that before subduction commences the isotope composition of what is to become the overriding mantle wedge is most likely to plot on the trend of the Group I samples (Figure 2). In a very small number of cases it might plot on a flat-lying trend similar to the Group II rocks.

SUBDUCTED OCEAN CRUST

An increasing volume of Nd- and Sr-isotope results are now available on the various components which comprise the ocean crust. The majority of unaltered basalts plot in the field for MORB at the low $^{87}Sr/^{86}Sr$ end of the field of Group I rocks in Figure 2. Ocean island, or 'hot spot', magmas are volumetrically less significant and they tend to have higher $^{87}Sr/^{86}Sr$ and lower $^{143}Nd/^{144}Nd$ ratios.

Hydrothermal alteration of ocean-floor rocks near mid-ocean ridges results in fractionation between Nd- and Sr-isotopes (Figure 3). The $^{87}Sr/^{86}Sr$ of the basalts is increased by interaction with seawater ($^{87}Sr/^{86}Sr = 0.7092$) whereas the $^{143}Nd/^{144}Nd$ ratio appears to be unaffected (O'Nions *et al.*, 1978). This reflects both the greater mobility of Sr and the extremely small quantities of Nd that are present in seawater (Høgdahl *et al.*, 1968). Moreover, since material that has been altered hydrothermally is likely to be more easily mobilized during subduction, we may predict that it will be most likely to influence the geochemistry of destructive plate margin magmas.

Results on deep-sea sediments and authigenic material such as Mn nodules indicate that the Nd-composition of ocean-water is about 0.5124 (O'Nions *et al.*, 1978; Hawkesworth and Elderfield, in preparation). Authigenic material, therefore, has Nd- and Sr-isotope compositions of 0.5124 and 0.7092, respectively, while sediments with more continental detritus tend to have higher $^{87}Sr/^{86}Sr$ and lower $^{143}Nd/^{144}Nd$ ratios (McCulloch and Wasserburg, 1978).

DESTRUCTIVE PLATE MARGIN MAGMAS

Combined Nd- and Sr-isotope studies have now been reported on destructive plate margin magmas from areas as different as the Marianas (DePaolo and Wasserburg, 1977), the South Sandwich Islands (Hawkesworth *et al.*, 1977), the South American Andes (Hawkesworth *et al.*, 1979a), and the Lesser Antilles (Hawkesworth *et al.*,

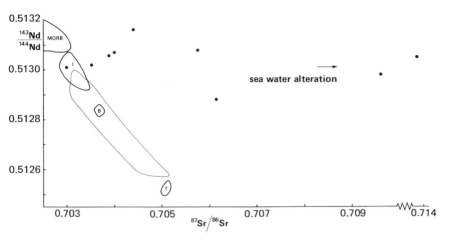

Figure 3 $^{143}Nd/^{144}Nd$ and $^{87}Sr/^{86}Sr$ ratios on ocean-floor basalts which have been altered by hydrothermal interaction with seawater (from O'Nions *et al.,* 1978), compared with the main trend of mantle-derived volcanic rocks (Group I, Figure 2). I = Iceland; B = Bouvet; T = Tristan da Cunha

1979c; Hawkesworth and Powell, 1979). These results are summarized in Figures 4 and 5, and the majority of them are displaced to high $^{87}Sr/^{86}Sr$ ratios compared to the main trend of the Group I rocks (Figure 2). Apart from N Chile, which will be discussed later, the Marianas would appear to provide the obvious exception to that rule (Figure 4). However, the available $^{87}Sr/^{86}Sr$ ratio on a back arc basalt in the Marianas is as low as 0.7026, so that further work is needed to assess whether these arc basalts also reflect some displacement to higher $^{87}Sr/^{86}Sr$ ratios (see Figure 1).

It is of interest that the displacement to higher $^{87}Sr/^{86}Sr$ ratios is observed in both *island arc tholeiites* and *continental andesites*. This suggests that, despite the marked differences in major element chemistry, similar processes have influenced the Sr-isotope geochemistry of magmas at both oceanic and continental destructive margins. In view of the results from ocean sediments and altered ocean-floor basalts (Figure 3), the relatively high $^{87}Sr/^{86}Sr$ ratios of these magmas are interpreted as evidence for a contribution from the subducted lithosphere. However, isotopes cannot discriminate between melting the slab and merely driving off aqueous fluids during dehydration.

Trace element geochemistry

It is beyond the scope of this contribution to review all aspects of the trace element geochemistry of destructive margin magmas. This section will therefore consider only evidence which bears on the problem of whether material from the subducted lithosphere is released by melting or by dehydration. Significantly, perhaps,

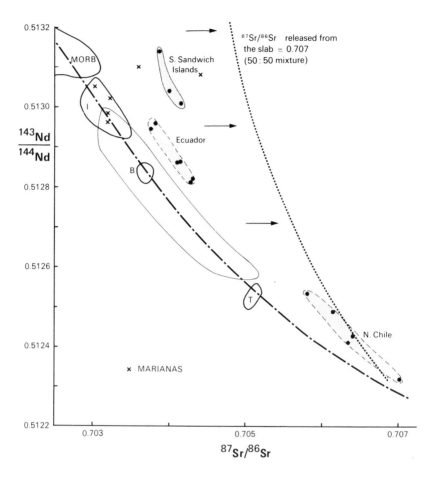

Figure 4 Nd- and Sr-isotopes in some destructive plate margin magmas compared with the main trend (Group I, Figure 2) of most mantle-derived volcanic rocks (for discussion see text)

geophysicists agree that the downgoing slab will dehydrate, but opinions differ on whether conditions ever rise above the solidus during steady-state subduction (Anderson *et al.*, 1978).

In the South Sandwich Islands, Hawkesworth *et al.* (1977) estimated that the parental magmas had trace element ratios of K/Rb = 300, K/Ba = 30, Rb/Sr = 0.08, Ce_N/Yb_N = 0.8; that is, they were relatively enriched in alkali and alkaline earth elements, but depleted in light rare earths. The displacement to high $^{87}Sr/^{86}Sr$ ratios (Figures 1 and 4) indicates that there is a contribution from the subducted lithosphere, but if partial melting were primarily responsible for the relative enrichment in alkaline elements it would also be expected to cause enrichment in light relative to heavy rare earths (i.e. $Ce_N/Yb_N \gg 1$). Since that is not observed, it is suggested that the fractionation between alkaline elements and the light rare earths is evidence that only the alkaline elements are released from the

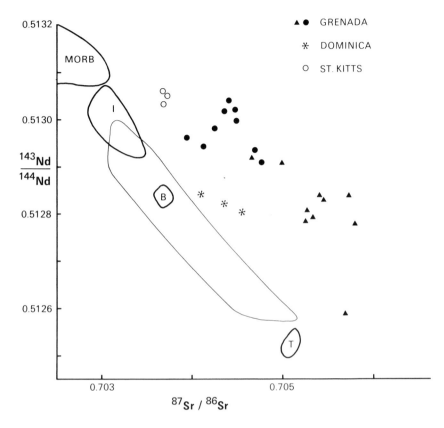

Figure 5 $^{143}Nd/^{144}Nd$ and $^{87}Sr/^{86}Sr$ ratios from volcanic rocks in the Lesser Antilles island arc compared with those from MORB and some Atlantic ocean islands. I = Iceland; B = Bouvet; T = Tristan da Cunha. Data from Hawkesworth *et al.* (1979c), Hawkesworth and Powell (1979), and O'Nions *et al.* (1977)

subducted slab and that this takes place not by melting but by dehydration.

Results from other island arcs are perhaps more ambiguous but they are certainly consistent with a model of alkaline enrichment in H_2O-rich fluids from the subducted lithosphere. On Grenada, Hawkesworth *et al.* (1979c) invoked such a model to reconcile their observation that the proposed ocean-floor 'finger-print' of relatively high $^{87}Sr/^{86}Sr$ ratios (Figure 5) is present in SiO_2-undersaturated basalts which are most unlikely to have been derived by melting oceanic crust. On a more general level, Pearce and Wright (in preparation) have demonstrated that high Sr/Ce ratios are such a common feature of island arc magmas that they may be used to distinguish them from mid-ocean ridge and intraplate basalts. Clearly, that is the trace element analogue to the relatively high $^{87}Sr/^{86}Sr$ ratios revealed by combined Nd- and Sr-isotope studies (Figures 4 and 5).

Along continental margins, such as the Andes (see Thorpe and Francis, this volume), the picture is even more complex. The magmatic rocks tend to be chemically more evolved than many island arc magmas, and there is little agreement about whether this reflects differences in source chemistry and/or extensive fractionation from basalt precursors. In addition, there is the risk of contamination with crustal material. However, despite these problems, it is possible to identify two important general features of continental margin magmas which are relevant to this discussion.

Firstly, many workers have pointed out that the degree of enrichment of some trace elements in calc-alkaline rocks was too great for them to be derived by simple partial melting of the subducted ocean crust. Lopez-Escobar *et al.* (1977) estimated that for some incompatible elements the Chilean andesites would represent < 5% partial melt of ocean-floor basalt, while on major element grounds they should represent at least 30% melt.

Secondly, the distribution of trace elements in calc-alkaline rocks (see Tarney and Saunders, this volume) differs significantly from that in intraplate magmas. This was emphasized by Hawkesworth *et al.* (1979a) in a comparison of the isotope and trace element geochemistry of intraplate lavas from Patagonia and calc-alkaline andesites from the Andean Cordillera. Their results are summarized in Figure 6, where it can be seen that elements as different chemically as Rb, Sr, Nb and Zr undergo broadly parallel enrichment in the Patagonian rocks, but that only the alkaline elements are enriched significantly in the calc-alkaline andesites.

In summary, both these features of the trace element geochemistry of continental margin andesites are consistent with the results from island arc rocks. They suggest that in the areas studied, the alkaline elements have been preferentially released from the subducted lithosphere by dehydration rather than by melting.

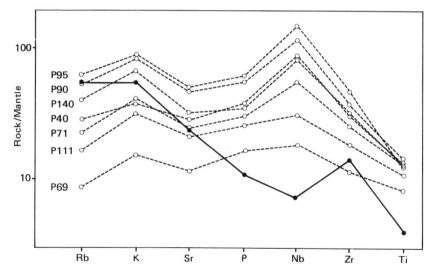

Figure 6 Mantle-normalized trace element distribution patterns for some Patagonian intra-plate lavas (open circles) compared with the average of available analyses on calc-alkaline andesites (56–58% SiO_2) from Ecuador and N Chile (solid line), after Hawkesworth *et al.* (1979a). (From *Earth and Planetary Science Letters,* 1979, vol. 42, p. 51, Figure 4a; by courtesy of Elsevier, Amsterdam)

For the purposes of this discussion we shall therefore adopt the dehydration model and briefly consider some of its consequences.

Dehydration of the subducted lithosphere

Combined Nd- and Sr-isotope results indicate that many destructive margin magmas contain a contribution from the subducted lithosphere. Trace element arguments suggest that the contribution is released during dehydration rather than in melting of the downgoing slab. Moreover, there is some evidence, e.g. from the South Sandwich Islands (Hawkesworth *et al.,* 1977), that the rare earth elements may not be released in significant amounts from the subducted ocean crust. If true, and it must be emphasized that this still remains to be tested, several points should be noted:

1. The Nd-isotope composition of destructive margin magmatic rocks, and its relationship to their Sm/Nd ratios, may be largely unaffected by the subduction process. In this case they will reflect the pre-subduction evolution of the overriding lithosphere.
2. The alkali and Sr-isotope geochemistry of destructive margin magmas will reflect both the pre-subduction history of the overriding lithosphere and a contribution from the subducted ocean crust.
3. If Sr, but not Nd, is released from the subducted slab then it is not possible to distinguish Sr derived by dehydration of altered basalt from that derived from subducted sediments: both will tend to increase the $^{87}Sr/^{86}Sr$ ratios of the destructive plate margin magmas (Figure 4).
4. The size of the displacement to higher $^{87}Sr/^{86}Sr$ ratios will depend on the amount and the composition of the strontium, both released from the slab and present in the overriding lithosphere.

This last point may be illustrated schematically. The dotted line on Figure 4 represents mixtures of equal amounts of Sr from the slab ($^{87}Sr/^{86}Sr = 0.707$, i.e. some mixture of basalt and seawater or sediments) and Sr from the overriding lithosphere which was assumed to lie on the main trend of the Group I samples (Figure 2). It emphasizes that the input of radiogenic Sr causes a significant displacement in areas like the South Sandwich Islands, but that if $^{87}Sr/^{86}Sr$ in the overriding lithosphere was high (perhaps 0.706–0.707, Figure 4) *before* subduction took place then, clearly, little or no displacement in $^{87}Sr/^{86}Sr$ will result.

Thus, one model for the different isotope composition of the andesites in Ecuador and N Chile (Francis *et al.,* 1977; Hawkesworth *et al.,* 1979a) is that they were different before subduction took place. This envisages that the variations in $^{143}Nd/^{144}Nd$ are analogous to those observed in ocean island and mid-ocean ridge rocks in the Atlantic (Figure 4) and that they reflect differences in the pre-subduction evolution of those portions of the upper mantle beneath South America. When radiogenic Sr is then released from the subducted lithosphere, the largest increases in $^{87}Sr/^{86}Sr$ may be expected in those areas which initially had low $^{87}Sr/^{86}Sr$ ratios, such as Ecuador.

Finally, the amount of displacement from the main trend of $^{143}Nd/^{144}Nd$ and $^{87}Sr/^{86}Sr$ will also depend on the amount of Sr present in the overriding lithosphere. Hawkesworth *et al.* (1979a) suggested that this might explain, at least in part, why the low-Sr rocks of the South Sandwich Islands are displaced twice as far from the main trend in Figure 4 as the higher-Sr rocks from Ecuador.

Crustal contamination

The continental crust is characterized by higher $^{87}Sr/^{86}Sr$ ratios than most of the upper mantle sampled by volcanism. Thus, high $^{87}Sr/^{86}Sr$ ratios in magmatic rocks are often taken as evidence that they have suffered contamination by crustal material *en route* to the surface. However, it is now known that $^{87}Sr/^{86}Sr$ ratios of 0.71 may be generated within the upper mantle (Erlank and Shimizu, 1977; Kramers, 1977; Hawkesworth and Vollmer, 1979), and that continental granulite-facies rocks may contain very unradiogenic Sr (Chapman, 1978). Arguments for or against contamination of magmas, particularly in the range of 0.705–0.708, based *solely* on Sr-isotopes are therefore much weaker than was believed previously, but it may be that further progress will result from combined Nd- and Sr-isotope studies.

Thorpe and Francis (this volume) have summarized the published $^{143}Nd/^{144}Nd$ and $^{87}Sr/^{86}Sr$ ratios on crustal rocks, and have outlined some general mixing relationships between crustal and mantle end-members. This section will therefore restrict itself to two more localized studies where combined Nd- and Sr-isotopes have been reported on magmatic rocks which are believed to have been contaminated by crustal material.

Norry *et al.* (1979) investigated the major, trace element and isotope geochemistry of a suite of lavas from the Gregory Rift in Kenya. The basalts have $^{143}Nd/^{144}Nd$ and $^{87}Sr/^{86}Sr$ ratios which plot in the main trend of the Group I samples (Figure 7), but a detailed study of the volcano Emuruangogolak suggested that the more evolved rocks had undergone *selective* contamination with radiogenic Sr. Emuruangogolak is of Quaternary age and has experienced repeated caldera collapse which indicates that a large magma chamber is present near the surface. Major and trace element studies (Weaver, 1976) show that the trachytes were probably produced from the basalts by low-pressure crystal settling, and yet the trachytes have much higher and more variable $^{87}Sr/^{86}Sr$ ratios than the basalts. The Nd-isotope composition of the two rock types appears to be the same within analytical error (see Figure 7), and it is noticeable that the trachytes have higher Nd but much lower Sr contents than the basalts. Norry *et al.* (1979), therefore, concluded that the Sr- but not the Nd-isotope ratios of the trachytes reflected contamination with continental crust.

The second example where interaction between mantle- and crustal-derived material appears to be well documented is in the Recent magmatic rocks of central Italy. In the more northerly Tuscan province the magmas are generally

86

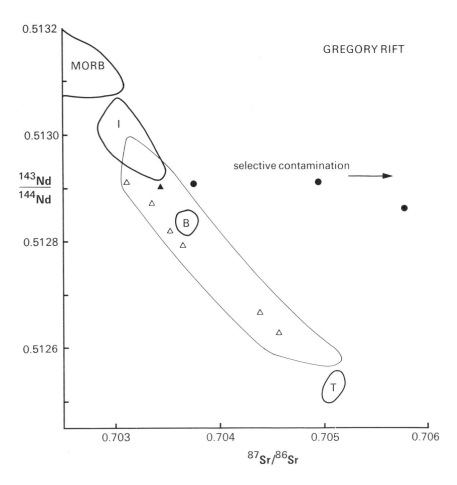

GREGORY RIFT

selective contamination

Figure 7 $^{143}Nd/^{144}Nd$ and $^{87}Sr/^{86}Sr$ ratios on recent basalts (triangles) and trachytes (circles) from the Gregory Rift (Norry *et al.*, 1979). Filled symbols denote rocks from the volcano Emuruangogolak

believed to reflect crustal anatexis, while Hawkesworth and Vollmer (1979) have argued that the typically SiO_2-undersaturated alkaline lavas south of Rome were derived from the upper mantle and that the primitive rocks have been largely unaffected by interaction with continental crust. In between, there is a transition zone where crust/mantle interaction has been demonstrated by Turi and Taylor (1976) who observed a striking correlation between increasing $\delta^{18}O$ and the degree of SiO_2 saturation. Although the transition zone has not itself been the subject of a detailed Nd- and Sr-isotope study, results are available from the areas of crustal- and mantle-derived rocks to the north and south, and these are summarized in Figure 8. Many of them have similar $^{143}Nd/^{144}Nd$ but very different $^{87}Sr/^{86}Sr$ ratios, so that in this area contamination of mantle-derived melts with continental crust will again result in a flat-lying trend on the Nd- and Sr-isotope diagram (Figure 8).

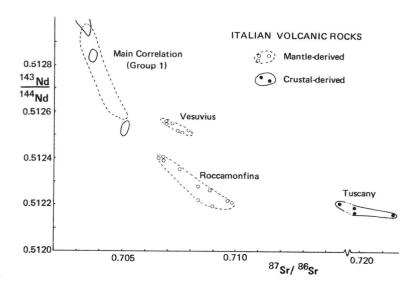

Figure 8 $^{143}Nd/^{144}Nd$ and $^{87}Sr/^{86}Sr$ ratios on recent mantle- and crustal-derived magmatic rocks in central Italy (Hawkesworth and Vollmer, 1979)

^{87}Rb decays about twice as fast as ^{147}Sm, and thus high Rb/Sr, low Sm/Nd continental crust will develop high $^{87}Sr/^{86}Sr$ ratios very much more rapidly than low $^{143}Nd/^{144}Nd$ ratios. In central Italy the oldest exposed basement is Hercynian and the crustal rocks have high $^{87}Sr/^{86}Sr$ ratios, but their Nd-isotopes are still similar to upper mantle rocks (Figure 8). By contrast, if the high $^{87}Sr/^{86}Sr$ ratios of the N Chilean andesites reflect crustal contamination, then the combination of Nd- and Sr-isotopes (Figure 4) indicates that the contaminant must have had low $^{143}Nd/^{144}Nd$ ratios which, in turn, implies that it was old rather than young continental crust. Whether contamination by old continental crust is likely in an area of such recent crustal growth as the Andes remains a matter for speculation (see also Thorpe and Francis, this volume).

Concluding remarks

The application of combined Nd- and Sr-isotopes to the study of destructive plate margin magmas has provided strong evidence that most of them contain a contribution from the subducted ocean crust mixed with material from the overriding lithosphere. Moreover, trace element arguments have been presented suggesting that in many instances this contribution may be released during dehydration rather than melting of the downgoing slab. Further work may enable us to estimate more precisely the chemical contribution from the subducted ocean crust, and to see back into the chemical 'events' which make up the pre-subduction evolution of the overriding lithosphere.

Finally, preliminary results suggest that combined Nd- and Sr-isotope investigations will be important in assessing the role of crustal contamination in continental magmatism. They will not provide a universally applicable yardstick by which contamination can be measured, but, when used in regional studies, they can place critical constraints on the nature of the proposed contaminant.

ACKNOWLEDGMENTS

I am grateful to Dr. J. Pearce for reviewing this paper, John Taylor for preparing the diagrams and Marilyn Leggett for typing the manuscript.

TRACE ELEMENT CONSTRAINTS ON THE ORIGIN
OF CORDILLERAN BATHOLITHS

J. TARNEY and A.D. SAUNDERS*
Department of Geological Sciences, University of Birmingham,
Birmingham B15 2TT, UK

Introduction

The problem of the origin of cordilleran batholiths cannot, because of their immense volume, be separated from the origin of the continental crust itself. Despite arguments as to when and how the crust was generated, the extent of continental reworking and mobilization and the importance of sediment recycling (see Fyfe, 1978, 1979; Brown, this volume), there can be little doubt from isotopic studies (Moorbath, 1977, 1978; McCulloch and Wasserburg, 1978; Hamilton et al., 1979) that large areas of continental crust represent new additions of sialic material from the mantle system at various periods. Tarney (1976) and Windley and Smith (1976) drew attention to the broad compositional similarities between Archaean continental crust and the cordilleran batholiths of North and South America. Tonalite is a dominant component in both − this is particularly so in the more deeply eroded parts of the Andean cordillera, where the plutonic rocks frequently have a distinct fabric and may even be foliated. Whether the crustal generation processes in the Archaean exactly paralleled those in the modern Andes, however, remains to be determined.

There is still no consensus as to the processes by which sialic material is generated, although there is a clear association of most, if not all, modern batholiths and andesitic volcanoes with zones of lithospheric plate subduction. Possible source regions for sialic magmas in subduction zones include the basaltic ocean crust (its composition modified, perhaps, by ridge hydrothermal activity), the mantle wedge above the subduction zone, ocean floor or continental margin sediments carried down the subduction zone, and melting of (or contamination with) continental crust during ascent of magmas. Of course all four might be involved, although in many cases isotopic evidence severely limits the amount of older continental material (either as crust or sediment) which can be involved in calc-alkaline magma genesis. Also very uncertain is the extent to which the range of magma compositions in cordilleran batholiths can be related by fractional crystallization and whether individual plutons represent partial melt compositions generated at source. Crucial in this respect is the nature of the deeper parts of batholiths, or the lower crust in general, which must contain a high proportion of cumulates if fractional crystallization is a dominant mechanism; evidence on the

*Present address: Department of Geology, Bedford College, Regent's Park, London NW1 4NS, UK

nature of the lower crust, however, is scant (Tarney and Windley, 1977). Many of the salient problems associated with calc-alkaline magma genesis have been summarized by Ringwood (1974, 1977) and Gill (1978). Ringwood's preferred model, for instance, involves partial melting of the basaltic (eclogitic) slab to form silicic liquids which react with the overlying mantle wedge to form garnet pyroxenite diapirs which eventually fractionate to produce calc-alkaline magmas.

In the following account we will attempt to constrain the petrogenetic processes associated with calc-alkaline magma generation by considering and comparing relevant trace element data on three different calc-alkaline (s.1.) plutonic provinces of different ages: (a) the Mesozoic–Tertiary batholith and associated volcanic rocks of the Antarctic Peninsula, which is typical of the Andean batholith; (b) a Proterozoic high-K plutonic complex from East Greenland; and (c) Archaean tonalitic-trondhjemitic orthogneisses from the Lewisian complex, NW Scotland. All conform to the I-type classification of Chappell and White (1974). In addition, to preface the discussion, we will briefly outline the trace element behaviour in the lavas of Deception volcano, Antarctic Peninsula, which range from olivine basalt to rhyodacite, and demonstrate the trace element relationships expected from fractional crystallization processes which lead to silicic liquids.

Trace element fractionation in Deception volcano

A detailed account of the geochemistry of the lavas of Deception volcano has been published by Tarney *et al.* (1977) and Weaver *et al.* (1979). The volcano is associated with the low-K calc-alkaline volcanic sequences of the S Shetland island arc, but is more directly linked with the initial stages of back-arc spreading in Bransfield Strait, separating the S Shetland Islands from the Antarctic Peninsula. Deception lavas range from olivine tholeiite to rhyodacite, although it is apparent from the low MgO (6%), Cr (140 ppm) and Ni (35 ppm) contents of the most primitive basalts that even these are considerably modified mantle melts (Table 1). The smooth consistent trends within the lava suite (Figure 1 and Weaver *et al.*, 1979) can best be explained by fractional crystallization. There is an initial Fe-enrichment until both Ti and Fe decrease sharply with titanomagnetite crystallization. Phosphorus also increases until apatite separates, while Sr consistently decreases because of early and continued plagioclase crystallization. Other incompatible elements such as Zr, Nb, K, Rb, Ba and even Na_2O increase progressively with fractionation, Na_2O reaching 7.5 wt. % in the dacites. There is a linear relationship between Zr and SiO_2. Na_2O/Zr and Ba/Zr ratios do, however, fall slightly, reflecting some entry of Na and Ba into plagioclase, but K and Rb increase systematically with increasing Zr, confirming that no potassic phase crystallized. Total rare earth element (REE) levels increase with fractionation, with only slight change in Ce/Yb ratio, and with the eventual development of distinct negative Eu anomalies in the dacites. Ce/Zr and Y/Zr ratios do reduce with fractionation, especially in the more silicic compositions, indicating increasing partitioning of REE into the major mineral phases as the magma becomes more silicic.

The range of trace element enrichments indicate at least 80% fractional crys-

91

Table 1 Analyses of Recent, Mesozoic, Proterozoic and Archaean calc-alkaline (s.l) rocks

	Deception Island Basalt B138.2	Rhyodacite P870.1	Antarctic Peninsula Gabbro TL587.1	Tonalite TL571.1	Granite D4613.1	Proterozoic plutonic complex, E Greenland Picrites N=4	Gabbros N=14	Diorites N=21	Granites N=33	Archaean gneisses Tonalites N=36	Trondhjemite 16V
SiO_2	51.89	68.02	52.76	58.99	76.04	47.0	51.7	57.1	71.8	67.62	68.38
TiO_2	1.49	0.55	0.61	0.83	0.16	0.94	1.26	1.63	0.26	0.38	0.23
Al_2O_3	16.20	14.99	15.65	16.24	12.70	7.7	15.5	14.1	14.5	15.61	16.39
tFe_2O_3	9.46	4.97	9.26	6.22	1.09	14.9	9.5	8.7	2.4	3.05	2.14
MnO	0.18	0.18	0.16	0.09	0.03	0.18	0.11	0.11	0.04	0.04	0.03
MgO	6.11	0.33	7.47	3.61	0.12	21.3	7.10	4.27	0.32	1.46	1.00
CaO	10.07	1.69	10.37	5.60	0.57	4.73	9.12	6.72	1.36	3.42	1.75
Na_2O	4.07	7.45	2.96	3.50	4.32	1.60	3.04	3.40	3.68	4.97	7.09
K_2O	0.28	1.69	0.76	2.56	4.53	0.72	0.89	2.30	5.08	1.13	0.78
P_2O_5	0.21	0.10	0.13	0.19	0.04	0.24	0.65	0.83	0.06	0.15	0.08
TOTAL	99.96	99.97	100.13	97.83	99.60	98.3	98.87	99.16	99.5	97.83	97.87

Trace elements (ppm)

Ni	35	2	41	6	<1	910	113	44	3	40	26
Cr	141	7	266	33	5	1500	160	60	7	26	13
La	8	28	13	22	19	35	77	130	120	25	10
Ce	21	64	24	36	49	60	160	250	230	49	21
Y	26	71	22	20	19	12	19	33	25	3	3
Zr	144	665	96	238	92	150	190	820	360	278	127
Nb	2	17	4	12	11	9	11	23	27	6	5
Rb	3	32	22	62	173	17	30	75	190	15	11
Sr	340	134	296	532	24	350	860	590	150	615	554
Ba	88	242	292	690	167	630	840	1490	830	984	609
Pb	4	12	9	9	22	16	16	19	26	14	21
Th	2	7	2	4	26	4	6	7	19	1.5	5
Zn	76	124	79	34	21	–	–	–	–	–	–

Note: Analyses of Deception Island lavas from Weaver *et al.* (1979); analyses of Antarctic Peninsula plutonic rocks from Saunders *et al.* (in press, b); analyses of E Greenland plutonic rocks from Wright *et al.* (1973); analyses of Archaean gneisses from Tarney *et al.* (1979). tFe_2O_3 = total iron as Fe_2O_3; N = number of analyses.

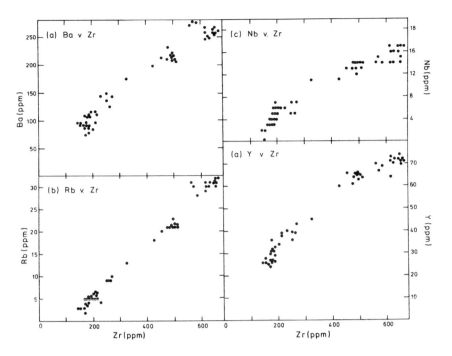

Figure 1 Trace element relationships in Deception basalt to rhyodacite lavas. Zirconium is used as a fractionation index

tallization. Notable is the strong coherence during fractionation of SiO_2, K_2O, Rb, Ba, Zr, Nb, Y and the REE.

Andean plutons of the Antarctic Peninsula

The Mesozoic history of the Antarctic Peninsula bears many resemblances to that of the Andean cordillera of South America. Indeed, before the opening of the Atlantic and the break-up of Gondwanaland, the two were continuous (de Wit, 1977) and formed a Pacific margin to the supercontinent. Mesozoic plutonic and volcanic rocks make up much of the exposed lithology of the Peninsula. By the late Mesozoic, however, the Antarctic Peninsula began to evolve on a different course. First, the opening of the Weddell Sea separated the Peninsula from the remainder of the Antarctic continent, and the locus of magmatic activity moved trenchwards. Second, during the Tertiary, sections of a spreading ridge in the SE Pacific were progressively consumed, from south to north, beneath the Peninsula, which appears to have led to a cessation of magmatic activity in the same direction. Third, the opening of Drake Passage in the mid-Tertiary finally led to the separation of the Antarctic Peninsula from S Chile and the development of the Scotia Sea. This culminated in the subduction of Atlantic ocean lithosphere beneath the region in the East Scotia Sea. The last few million years has seen the end of

93

Pacific lithosphere subduction at the S Shetland trench. Since the Mesozoic, in fact, there has been progressive fragmentation and oceanization of the region. Volcanic activity has gradually become dominant over plutonic activity and the volcanic products themselves have become distinctly more basic. This is the reverse of the normal evolutionary sequence of an island arc.

During the Mesozoic there was widespread volcanic and plutonic activity, the main features of which have been described by Weaver *et al.* (in press) and Saunders *et al.* (in press, a). Among the volcanic rocks there is a distinct transverse compositional variation. Those lavas erupted nearer the trench (on the S Shetland Islands) are dominantly basalts and basaltic andesites, those on the west coast of the Peninsula include a higher proportion of andesites, whereas

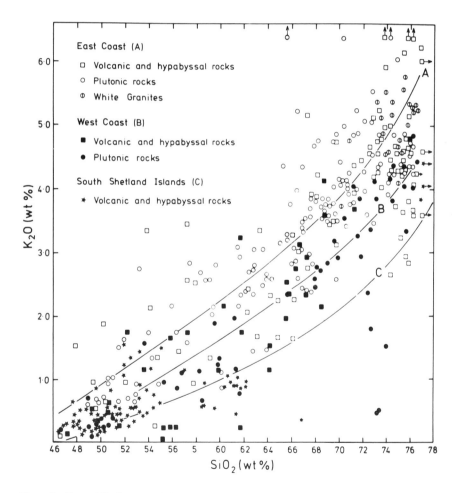

Figure 2 Plot of K_2O versus SiO_2 for Antarctic Peninsula Mesozoic volcanic and plutonic rocks, illustrating K-h variation across the Peninsula. Curves represent mean K_2O–SiO_2 trend for each region: A = east coast; B = west coast; C = South Shetland Islands. (From Saunders *et al.*, in press, b)

those on the east coast are dominantly rhyolitic. Superimposed on this is a 'K-h' variation, i.e. in lavas of equivalent SiO_2 percentage, K_2O increases progressively towards the east coast. Broadly the same variation is apparent in the plutonic rocks which, although not necessarily co-magmatic with the lavas, display a similar range of composition. Figure 2 illustrates this feature in both plutonic and volcanic rocks; note, however, that there is considerable dispersion at any given value of SiO_2. This may result in part from the fact that we have grouped together all Mesozoic igneous rocks with a broad range of absolute ages; but, more likely, it reflects the complex petrogenetic processes attending calc-alkaline magma genesis. That the K-h variation is spatially rather than temporally related is demonstrated by the fact that volcanic rocks on the S Shetland Islands, which were erupted in the Mesozoic, the early Tertiary, the late Tertiary and Quaternary, are all mainly low-K basalts and basaltic andesites.

Saunders *et al.* (in press, b) have also examined which other elements exhibit similar transverse variations to K_2O. It might be expected, for instance, that the feature might be shown by all incompatible elements. However, it is specifically only the strongly lithophile (LIL) elements K, Rb, Th, and to a lesser extent Ba, La and Ce (and we suspect U and Cs also), which show transverse variations. These are the elements that normally increase systematically with SiO_2 in calc-alkaline suites. Ce_N/Yb_N ratios also increase away from the trench. Other high field strength (HFS) trace elements such as Zr, Hf, Ta, Nb, Ti and P, which are normally incompatible in tholeiitic and alkali basalt suites, do not show any systematic transverse variation. Nor, incidentally, do they increase systematically with increasing SiO_2 in calc-alkaline suites. Zr, for instance, increases in abundance until about 60–65% SiO_2, then decreases sharply (Figure 3); Nb increases slowly in basaltic rocks, but then remains relatively constant in intermediate to acid compositions. This behaviour seems to be characteristic of many calc-alkaline provinces, but contrasts with the behaviour of Zr and Nb in Deception volcano (Tarney *et al.*, 1977; Weaver *et al.*, 1979), both of which increase linearly with SiO_2 in the compositional range olivine basalt to rhyodacite, a range of variation which is fully in accord with a fractional crystallization model.

This difference in behaviour of Zr, Hf, Nb, Ta etc. suggests that, if fractional crystallization is the main mechanism relating magma compositions in calc-alkaline suites, then some mineral phase (or phases) is removing HFS elements. Ilmenite, rutile, sphene, zircon and apatite are obvious contenders, but extensive cumulates rich in these minerals in calc-alkaline provinces are lacking. Indeed, most recognizable cumulate rocks actually have low abundances of HFS trace elements. In the authors' experience in fact, rocks which can be unequivocally classified as cumulates, even in the deeply eroded Patagonian batholith, form only a very small proportion of the batholith.

An alternative explanation is that HFS elements are retained in the source regions during generation of calc-alkaline magmas: this implies, of course, that more of the compositional variation in calc-alkaline suites is attributable to partial melting processes than to fractional crystallization. All volcanic and plutonic calc-alkaline rocks in the Antarctic Peninsula have high LIL/HFS element ratios compared with normal tholeiitic and alkaline rocks. This applies to the

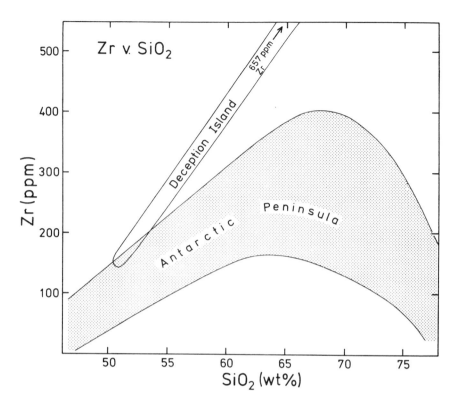

Figure 3 Plot of Zr versus SiO_2 for Antarctic Peninsula Mesozoic volcanic and plutonic rocks. Following an initial increase, Zr begins to decrease in intermediate to silicic compositions, contrasting with the $Zr–SiO_2$ relationship in Deception lavas. There is no significant difference in $Zr–SiO_2$ correlation across the Peninsula. (From Saunders *et al.*, in press, b; and Weaver *et al.*, 1979)

more mafic as well as to the siliceous compositions. The feature is also apparent in calc-alkaline and tholeiitic series lavas from Japan (Wood *et al.*, in press) and from primitive W Pacific island arcs (Mattey *et al.*, in preparation); indeed, it appears to be a fundamental characteristic of subduction zone magmas. There are two possible explanations: (a) that those LIL elements that are commonly mobile during low temperature hydrothermal alteration of ocean-floor basalts are carried into the source regions of calc-alkaline magmas during dehydration of the slab; and (b) that HFS elements are retained in the source regions by minor phases such as ilmenite, sphene, rutile, zircon, apatite, etc. which have enhanced stability under hydrous melting conditions. Both explanations may be valid and are not mutually exclusive (Saunders *et al.*, in press, b).

Rare-earth element distributions in Antarctic plutonic and volcanic rocks (Weaver *et al.*, 1979; Saunders *et al.*, in press, a; Tarney *et al.*, in press) are similar to those for Andean equivalents (Thorpe *et al.*, 1976; Lopez-Escobar *et al.*, 1976, 1977; Stern and Stroup, in press). Chondrite-normalized Ce_N/Y_N (or Ce_N/Yb_N) ratios are moderately low, and vary from about 2.0 close to the trench reaching up to about 8.0 on the east coast of the Peninsula, although there is moderate dispersion in these values in each area. The more silicic rocks usually have prominent negative Eu anomalies. In general, the Ce_N/Yb_N ratios are too low for the magmas to have been generated by equilibrium partial fusion of eclogite in the subducting slab (Gill, 1974), thus, either the mafic crust in the slab had not converted to eclogite or, more likely, the REE patterns reflect the source characteristics of the mantle wedge above the slab.

Proterozoic plutonic complex from East Greenland

A Proterozoic post-orogenic calc-alkaline complex was mapped and described by Wright *et al.* (1973) just north of the Nagssugtoqidian front near Angmagssalik, E Greenland. The complex forms an oval-shaped outcrop, *ca* 30 km in diameter, sharply cutting the Archaean grey gneisses and metasediments which had earlier been reworked during the Nagssugtoqidian deformation. Its occurrence near the major tectonic boundary of the Nagssugtoqidian with the older pre-Ketilidian Archaean craton of southern Greenland might suggest some link with subduction, but of course the exact tectonic relationships at the time of magma generation are unknown, and comparisons could equally well be drawn with some intraplate Pan-African or Proterozoic complexes (Gass, 1977; Emslie, 1978).

Nevertheless, the plutonic complex displays many features typical of Andean batholiths. All intrusive contacts are sharp and there are no pegmatites. The intrusive sequence, judged from cross-cutting relationships and xenoliths, begins with olivine gabbro and then successive magma compositions become progressively more leucocratic until the emplacement of K_2O- and Rb-rich pink microgranites. Picritic dykes cut the olivine gabbros, and appear to have been intruded early in the sequence. Some minor intrusions of microdiorite were emplaced after the main granitic bodies. Individual plutons are generally less than $1-2$ km in diameter, and there is no obvious zoning either within each pluton or in the complex as a whole.

Geochemically, this is a high-K plutonic complex (Table 1). The granitic plutons, the diorites, the gabbros and even the picrites have relatively high levels of K_2O, Rb, Ba, Sr, La, Ce and other LIL elements — much higher, in fact, than most of the Antarctic Peninsula and Andean plutons we have studied. Even HFS elements such as Ti, Nb, Zr and P are higher than in equivalent modern Pacific margin plutons. Nevertheless, the rocks still have high LIL/HFS element ratios like their Andean counterparts.

A detailed account of the petrology and geochemistry of this complex will be published elsewhere (Tarney *et al.*, in preparation); only directly relevant geochemical features will be discussed here. Mean REE patterns of the picrites, gabbros, diorites and granites are shown in Figure 4. The significant point is that the REE patterns of the four groups are essentially parallel and that all have high Ce_N/Yb_N ratios. The picrites have no Eu anomalies, the gabbros have both small positive and small negative Eu anomalies, the diorites tend to have small negative Eu anomalies, whereas the granites have distinct to very prominent negative Eu

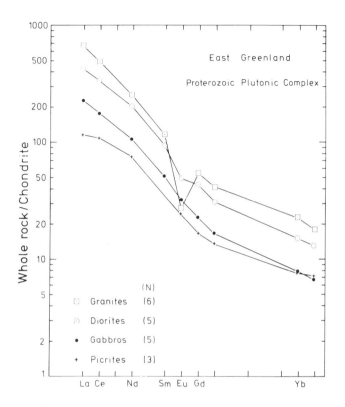

Figure 4 Mean chondrite-normalized rare earth patterns of picrites (3), gabbros (5), diorites (5) and granites (6) from the Proterozoic high-K plutonic complex, E Greenland. (From Tarney *et al.,* in preparation)

anomalies. Total REE concentrations are high and increase progressively between mean picrite and mean granite.

There is no doubt, from their distinctive REE and other trace element characteristics, that the whole range of rock types from picrite to pink granite are consanguineous. Evidence for fractional crystallization is provided by cumulate xenoliths within the gabbros, but most of the rock types appear to represent liquid compositions. Some caution in accepting a simple fractional crystallization model is necessary on account of the fact that the enrichment in some LIL elements (e.g. K_2O, Rb) from picrite to granite is greater than would be expected from the increase in REE levels. Moreover, although there is a *mean* increase in REE concentrations, in detail there is considerable overlap between the various rock types: REE abundances in the granites, for instance, cover the whole range.

Nevertheless, the fact that the picrites and olivine-gabbros have the same geochemical characteristics as the much more abundant diorites and granites suggests that mantle fusion is somehow involved in the genesis of these rocks. It would be difficult to conceive of a mechanism whereby these picritic and mafic magmas could have been derived by partial fusion of subducted ocean crust or lower crust. Whether the associated country rock gneisses are involved in the genesis of the granitic rocks is difficult to say in the absence of isotopic data. It is certainly possible but, if so, they must represent only a small degree of partial melting of such rocks because the heavy REE concentrations in the associated Archaean gneisses are usually low (Tarney and Windley, 1977). The problem is similar to that connected with the origin of Skye granites (Thorpe et al., 1977) which also have much higher HREE abundances than Lewisian country rocks.

The bronzite picrite dykes do not have cumulate characteristics and must represent fairly high degree mantle melts. They are similar, petrographically and chemically, to the 2200 Ma bronzite picrite dykes from the Lewisian (Tarney, 1973). It follows that, using arguments similar to Sun and Nesbitt (1977), their chemistry reflects that of their mantle source. Thus, the latter must have had 'enriched' geochemical characteristics, i.e. high contents of LIL and HFS elements, and high REE levels with fractionated Ce_N/Yb_N ratios. This, of course, is a characteristic of the whole plutonic suite, picrite to pink granite, and strongly implies that these geochemical characteristics are somehow mantle-derived. There would seem to be no necessity to directly involve subducted ocean crust in their genesis.

Archaean Lewisian orthogneisses

The geochemical characteristics of Lewisian orthogneisses have been described by Tarney et al. (1972), Sheraton et al. (1973) and Tarney (1976), and their REE chemistry in particular by Drury (1978), Tarney et al. (1979) and Weaver and Tarney (in preparation). The gneisses range in composition from ultrabasic through basic and intermediate to acid, and produce a smooth 'calc-alkaline' trend on an AFM diagram (Sheraton et al., 1973). This smooth trend, although well defined, may in part be an artefact of tectonic and metasomatic processes (Beach and Tarney, 1978; Tarney et al., 1979) since compositional distributions in many other Archaean gneiss terrains are strongly bimodal (Barker and Peterman, 1974; Weaver et al., 1978). The dominant component, as in Andean batholiths, is tonalite; but the more silicic differentiates, especially in the granulite terrains, are trondhjemitic rather than granitic and have low K_2O and Rb contents and very high Na/K and K/Rb ratios, though these features have been enhanced by the expulsion of K and Rb during the granulite facies event.

The main point of interest is why, starting with the same dominant magma component (tonalite), the fractionation trend is toward highly sodic siliceous trondhjemites rather than to the more usual potassic granite? Isotopic evidence indicates that these gneisses were derived from the mantle system by igneous processes 2900 Ma ago (Hamilton et al., 1979), just as isotopic data on Andean batholiths also indicate essentially mantle derivation in more recent times (cf. Brown, 1977). Clearly, the mechanisms or the conditions of magma generation and fractionation must have been different. This is reflected also in the trace element characteristics of the gneisses (Tarney et al., 1979). For instance, the gneisses are much more calcic at equivalent SiO_2 levels and Sr remains high ($\sim$ 500 ppm) even in quite siliceous compositions, contrasting with the more normal decrease in Sr over the same compositional range in Andean batholiths.

The behaviour of the rare earth elements is also very different. Instead of the general increase in REE concentrations with the gradual development of negative Eu anomalies in more silicic compositions seen in the Andean and E Greenland calc-alkaline plutons, the Lewisian REE distributions show the following *general* features: (a) the mafic gneisses have flat to moderately REE-enriched patterns with no or slight negative Eu anomalies; (b) the intermediate gneisses have fractionated light REE-enriched patterns; (c) the tonalitic gneisses have strongly fractionated REE patterns showing marked heavy REE depletion and generally positive Eu anomalies; and (d) the most siliceous tonalites and trondhjemites have very strongly fractionated REE patterns showing extreme heavy REE depletion and most have prominent positive Eu anomalies (Figure 5). These features are not unique to the Lewisian, but have been described also in a Proterozoic gabbro–diorite–tonalite–trondhjemite complex from SW Finland (Arth *et al.*, 1978). Although Archaean plutons and orthogneisses tend to have much more fractionated REE patterns than Andean equivalents (Arth and Hanson, 1975; Tarney and Windley, 1977; Compton, 1978), plutons with essentially Andean REE characteristics are also known from the Archaean (Hunter *et al.*, 1978). Likewise, some Andean and Phanerozoic plutons have quite strongly fractionated REE patterns (Buma *et al.*, 1971; Frey *et al.*, 1978; Stern and Stroup, in press). Nevertheless, some differences between Archaean and modern crustal generation processes are indicated.

Arth *et al.* (1978) considered that hornblende fractionation could account for the pattern of REE distributions in the Finnish trondhjemites, the important feature of hornblende being the high partition coefficients for HREE and the fact that hornblende develops a negative Eu anomaly in equilibrium with silicic liquids (Arth and Barker, 1976). Thus, extensive hornblende fractionation should yield residual liquids with HREE depletion and positive Eu anomalies. Arth *et al.* did not, however, exclude partial melting mechanisms with residual hornblende control. Garnet is also potentially capable of reproducing the same features and Tarney *et al.* (1979) suggested that either hornblende fractionation or partial melting of a garnet amphibolite mafic source at high water pressures was capable of accounting for the geochemical features of Lewisian gneisses. A partial melting model was favoured on account of the bimodal characteristics of many Archaean gneiss terrains and the lack of obvious cumulates. If high water pressures are an essential factor in the mechanism of development of trondhjemite liquids with positive Eu anomalies, it follows that such liquids may be restricted to deeper crustal levels. They may, of course, be subsequently transformed to dry granulite-facies assemblages through changes in fluid composition (Tarney and Windley, 1977; Hamilton *et al.*, 1979). Dacitic residual liquids, with high Na/K ratios and major element compositions similar to trondhjemites, may be produced by high-level crystal fractionation, as in Deception volcano (Weaver *et al.*, 1979). However, such liquids have much higher REE contents and negative Eu anomalies, and were derived from parental basaltic magmas with high Na/K ratios.

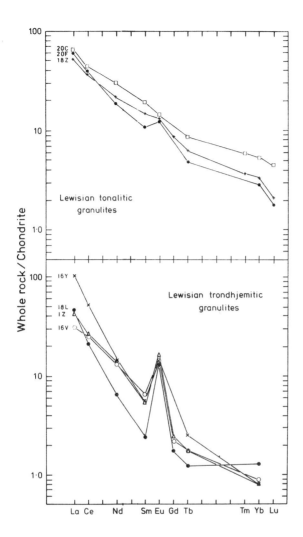

Figure 5 Typical rare earth patterns of Lewisian tonalites and trondhjemites. (From Tarney *et al..* 1979; and Weaver and Tarney, in preparation)

Discussion

It is clear from the trace element relationships outlined above that there are many different fractionation paths between basic magmas and granite (s.1). This may imply an equal diversity of P, T, P_{H_2O} and P_{O_2} conditions controlling magma genesis and subsequent crystal fractionation. Additionally, there are variations in source compositions to consider. Major element relationships, as portrayed for instance on AFM diagrams, convey a deceptively simple picture of the petrogenetic processes involved and are of limited value in elucidating these processes.

The major problems connected with the origin of batholiths are: (a) the extent to which the compositional variations can be accommodated by established fractional crystallization processes and whether they result from partial fusion; (b) the ultimate source of calc-alkaline magmas — the subducted ocean crust, the overlying mantle wedge or the continental crust itself; and (c) the conditions under which calc-alkaline magmas are generated. We briefly consider these problems below, in the light of the trace element data.

There is no doubt that crystal fractionation of basic magmas can yield silicic liquids. On Iceland, tholeiitic magmas evolve through ferro-basalts to produce small quantities of rhyolite or granophyre, and the trace element relationships are perfectly consistent with fractional crystallization of anhydrous mineral assemblages (Wood, 1978). Parallel relationships are seen in some ophiolite complexes (Saunders *et al.,* 1979) where there are moderate quantities of plagio-granite and extensive gabbro cumulates. Deception lavas also fractionate to rhyodacite, but the trend towards siliceous compositions sets in at an early stage, without the development of ferro-basalts, and intermediate andesitic lavas are more prominent. In all three cases the granitic liquids produced have higher levels of REE and other incompatible elements (including HFS elements) than their basaltic parent magmas, and their REE patterns have distinct negative Eu anomalies. There is no significant fractionation of REE distributions during the generation of such granite liquids, and in general REE, LIL and HFS elements behave in a coherent fashion and maintain consistent ratios.

That fractional crystallization is the main mechanism relating the range of magma compositions in cordilleran batholiths is much more difficult to establish. General arguments against include: (a) the small proportion of potentially parental gabbros and diorites in relation to the very large volume of granitic rocks; (b) the small proportion of recognizable cumulates in relation to the large volume implied by the liquid fractionation trends; and (c) the high viscosities of silicic magmas which are not conducive to crystal settling. With regard to the latter point it may be noted that the dominant rock type in cordilleran belts, tonalite, frequently has a high proportion of mafic xenoliths scattered throughout which show little evidence of gravitative settling, yet in theory these should have a settling velocity several orders of magnitude higher than that of individual crystals. On a more specific level, too, there are ambiguities in the trace element evidence. The strongly lithophile elements, K, Rb, Cs, Th and U, invariably correlate well with SiO_2, which might imply a fractional crystallization relationship; but since these elements are among the most incompatible trace elements, they would not be significantly fractionated by partial melting processes either. Most notable is the marked decoupling in the behaviour of different groups of trace elements such as the REE, the LIL elements (K, Rb, Ba) and the HFS elements (Zr, Nb) in calc-alkaline plutonic complexes, which contrasts with the much more coherent behaviour of these groups in the examples of fractional crystallization discussed above and in basaltic systems generally (cf. Tarney *et al.,* 1978). As an example, Figure 6 shows a plot of Nb versus Rb (two of the most incompatible trace elements) for rocks from the E Greenland complex. In the picritic and gabbroic rocks the abundance of both varies by a factor of three, but the Rb/Nb ratio remains close to 1.6. However, all the more siliceous plutonic rocks have much

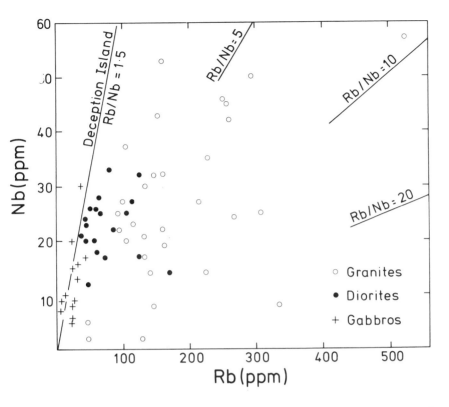

Figure 6 Plot of Nb versus Rb for plutonic rocks from the Proterozoic intrusive complex, E Greenland

higher, but very variable, Rb/Nb ratios (2—20) and much higher Rb concentrations. (By contrast *all* the Deception lavas, basic and siliceous, have Rb/Nb ratios close to 1.5). It is difficult to account for the lack of Nb in the more fractionated siliceous rock types. Nb does not correlate significantly with any of the other major trace elements, hence there is no obvious indication of a mineral phase (e.g. sphene, ilmenite) which might selectively remove Nb from the melt. Antarctic Peninsula plutons show similar features, but they are not so enriched in Nb or Rb as the East Greenland complex. Such trace element relationships caution against too readily accepting fractional crystallization as the main mechanism relating rock types in cordilleran batholiths. An alternative explanation is that each individual pluton has its trace element characteristics determined by the partial melting conditions and by the relative stability and nature of minor mineral phases in its source region. Successive plutons would have related, but different, geochemistries depending upon the thermal and P_{H_2O} conditions of magma generation. The main obstacle to such a model is the problem of deriving siliceous magmas with high LIL element abundances from an ultramafic mantle or subducted basaltic crust with relatively low LIL element abundances. Experimental studies suggest that hydrous melting of peridotite might yield melts as siliceous as basaltic andesite

103

(Kushiro, 1972) while wet melting of eclogite could yield dacitic magmas (Lambert and Wyllie, 1972).

Thermal models of subduction zones (Anderson *et al.,* 1978) indicate that, under the present geothermal regime, the downgoing slab may dehydrate, but will not necessarily melt. The REE data on Andean andesites (Thorpe *et al.,* 1976; Lopez-Escobar *et al.,* 1977) and on Antarctic plutons favour mantle derivation, as does the occurrence of picrites and olivine gabbros in the E Greenland complex. As we have argued, dewatering of the slab may carry selected LIL elements into the overriding mantle wedge and may contribute to the high LIL/HFS element ratios in any resultant magmas. Melting is, however, constrained by the energy needed to overcome the latent heat of fusion and this is provided by the heat capacity released by the lowering of the mantle solidus under hydrous conditions. In general, therefore, degrees of melting are likely to be low, except perhaps during the initial stages of subduction or in regions of active back-arc spreading where hot mantle is rising behind the arc. Silica transfer in supercritical fluids from the downgoing slab (quartz eclogite) into the overriding mantle wedge may also occur – indeed quartz veins are abundant in some glaucophane schist terrains such as on the Ile de Groix, Brittany, and point to the extreme mobility of silica in subduction zones. This may significantly affect the composition of the overlying mantle wedge and permit the generation of more siliceous (and more LIL element-enriched) magmas than would otherwise be possible. Although such a process may reduce the requirement for high-level crystal fractionation, it would not necessarily eliminate it.

The occurrence of picrites and olivine gabbros within the E Greenland complex not only indicates mantle involvement in the petrogenesis of the calc-alkaline complex, but the high contents of Ba (600 ppm), Sr (400 ppm), Rb (20 ppm), La (30 ppm), Zr (170 ppm) and Nb (10 ppm) in the picrites also suggest that the mantle source may be significantly enriched in incompatible elements. Because the picrites represent high degrees of mantle melting, it follows that the lower degrees of (hydrous) melting necessary to produce andesitic or tonalitic compositions would lead to much greater enrichment in incompatible trace elements. This is certainly observed for the LIL elements, but not for HFS elements such as Nb. HFS elements may be retained in the mantle source by minor phases such as ilmenite or rutile under hydrous melting conditions. In this model the high LIL element contents are derived directly from the mantle source (in much the same way as alkali basalts have high LIL element contents), without the necessity of a LIL component from the downgoing slab. The model does imply, however, that the subcontinental mantle sampled during calc-alkaline magma genesis is enriched in incompatible elements, perhaps by earlier metasomatic processes (Lloyd and Bailey, 1975) or by veining with enriched undersaturated liquids (Frey and Green, 1974; Tarney *et al.,* 1979) derived by incipient mantle melting at greater depths. Such veins are more likely to contain mineral phases such as ilmenite, rutile, zircon and apatite necessary to retain HFS trace elements during hydrous melting; indeed, these phases are commonly observed in enriched mantle nodules. Differences in the composition of the subcontinental and the suboceanic mantle may, to a first approximation, explain the generally higher lithophile element abundances in continental margin calc-alkaline igneous rocks.

The main geochemical differences between the tonalitic gneisses of Archaean high-grade terrains, such as the Lewisian and the tonalitic plutons of modern cordilleran batholiths, concern the generally higher Na/K ratios and moderate to extreme HREE depletion in Archaean gneisses together with the tendency to develop positive Eu anomalies rather than negative Eu anomalies in the residual liquids. It is important to bear in mind that not all Archaean plutonic rocks show these features, although HREE depletion is more common in Archaean tonalites. In many other respects, such as the levels of Sr, Ba, LREE, Zr, Nb etc., there are no significant differences. Assuming that the production of tonalitic magmas is related to subduction, it is possible to ascribe this difference to a component derived by hydrous melting of the downgoing slab. Partial melting of an eclogite or garnet amphibolite mafic crust at high water pressures, leaving residual garnet and/or hornblende, would produce the required HREE depletion and positive Eu anomalies in the generated liquids. These would have high Na/K ratios in the oceanic crust, and the high water pressures would ensure hornblende-dominated fractionation leading to trondhjemitic liquids with positive Eu anomalies (Tarney *et al.*, 1979). Such hydrous liquids would not be able to migrate to high crustal levels (Harris *et al.*, 1970) but tonalitic rocks with these geochemical characteristics might make up a significant proportion of the Archaean lower crust.

Differences between Archaean and modern crustal generation processes can thus be ascribed to an increased component due to melting of the downgoing slab in the early Precambrian. This results from the higher geothermal gradients in the Archaean; in modern cordilleran belts the evidence points to the fact that the downgoing slab does not melt significantly but may contribute LIL elements and silica to the source regions in the overriding lithosphere during dehydration. This may also have occurred in the Archaean, and may have given rise to magma types similar to those in modern cordilleran belts. Although Green (1975) has argued that geothermal gradients in the Archaean may have been too high to permit the conversion of oceanic crust to eclogite in the downgoing slab, other arguments suggest that the geothermal gradient might not be so extreme (Tarney and Windley, 1977; Weaver and Tarney, 1979).

ACKNOWLEDGMENTS

We should like to thank Giselle F. Marriner and Barry L. Weaver for the REE data on which Figures 4 and 5 are based. Studies in the Antarctic Peninsula are supported by the Natural Environment Research Council.

THE CHANGING PATTERN OF BATHOLITH EMPLACEMENT DURING EARTH HISTORY

G.C. BROWN

Department of Earth Sciences, The Open University,
Walton Hall, Milton Keynes MK7 6AA, UK

Introduction

Granite batholiths of all ages comprise an endless permutation of coarse-grained crystalline rocks based on the mineralogical theme feldspar ± quartz ± micas ± amphiboles. It is usual to find gabbro-diorite-tonalite-adamellite (sometimes trondhjemite) sequences, with a total range in silica from 45 to 78%, such as those described earlier from Peru (Pitcher, this volume). Tonalite is the dominant rock type in the continental crust and, despite near-surface metamorphic and sedimentary processes that bring about chemical segregation, the average composition of the continents is virtually tonalite (cf. Brown and Hennessy, 1978). Although calc-alkaline tonalite-dominated batholith suites occur from Archaean (e.g. Barker and Arth, 1976) to recent times (Bateman and Dodge, 1970; Pitcher, 1978), there are widespread alkaline and peralkaline, usually anorogenic suites, particularly from the Proterozoic (e.g. Emslie, 1978). These two suites are chemically distinct and yet, in many cases, alkaline granites are sufficiently well developed to have been termed 'batholithic'. They may, therefore, constitute an important contribution to past crustal growth. One of the two principal objectives of this paper is to discuss the relationship of the two major granite suites (calc-alkaline and alkaline) to tectonic conditions and magmagenetic controls. It will be argued that both suites have occurred throughout geological time, but with varying relative abundances; and that calc-alkaline magmatism has predominated, but that the importance of alkaline magmatism in crustal growth has been under-emphasized.

However, in discussing granite plutonism and Earth history it is also important to establish the causes and rates of modern crustal growth as a basis for comparison. Some interesting side-issues about continental growth rates throughout geological time stem from this discussion and these bear on recent disagreements in the literature. Thus, in addition to the geochemical/tectonic pattern of *crustal growth,* the second objective of this paper is to review the *role* of granites *(sensu lato)* and their evolution rate with time.

Crustal growth rates, past and present

Thirty years ago this volume would have debated the origin of batholiths from magmas or by *in-situ* sediment transformation (cf. H.H. Read's classic work of

1957). Fifteen years ago agreement had been reached on this issue, but it would have been heresy to suggest that granite magmas are generated other than in the crust. Today, there is remarkable agreement that most modern voluminous granite magmas are formed in regions of oceanic plate destruction and thus are dominated by subcrustal processes. This means that the continental crust receives a continuous supply of new material and the 'Ringwood cycle' (Ringwood, 1974), involving a two-stage crustal refinement from the mantle by ridge and subduction zone melting, provides a means of viewing this process. So the scale of the debate has shifted and today we are concerned with describing the extent of crust–mantle interaction above subduction zones (see, for example, Hawkesworth et al., 1979a; Hawkesworth, this volume; Thorpe and Francis, this volume) and the past variations in granite-forming processes.

The present writer takes the view (cf. Brown, 1977; Brown and Hennessy, 1978) that modern calc-alkaline batholiths are furnished by between, say, a 70 and 100% contribution from melts derived from the upper mantle above subduction zones, with the remainder coming from the lower crust. To judge from the geochemical and isotopic time and space variations in batholiths, it seems that the crustal contribution is proportional to the thickness of the crustal pile which grows with *time* in active zones (cf. Thorpe and Francis, 1979). For example, a K-h trend is often recognized (Best, 1975) and a K-time trend has been beautifully documented by recent Peruvian studies (Pitcher, 1978). Analogous initial $^{87}Sr/^{86}Sr$-time trends are known from N Chile (McNutt et al., 1975) but may not occur in Peru (Atherton et al., this volume). Initial ratio-space variations are better documented and spectacular results by Armstrong et al. (1977) testify to the involvement of crust in producing the Oregon–Washington–Idaho batholith. A sharply defined boundary separates those granites which rose through a Phanerozoic eugeosynclinal pile ($^{87}Sr/^{86}Sr$ initial = 0.702–0.704) from those (0.706–0.707) that rose through old Proterozoic basement. These are not crustal ratios, but they indicate a component of crustal strontium which is most likely to have been incorporated by mixing of ascending mantle magma with a bordering crustal melt zone (cf. Hodge, 1974).

Whether we take a 70 or 100% mantle contribution, the fact remains that new mantle-derived magmas (intrusive and extrusive) arrive in the crust at modern rates approaching, but not exceeding, 0.5 km^3 a^{-1} (Francis and Rundle, 1976; Brown, 1977). This is the modern rate of continental *accretion* and it raises the question of whether the continents are also growing at this rate. The problem has been discussed by Fyfe (1978, 1979) from whose work Figure 1 is adapted.

Model 1 is a steady-state view of crustal growth whereby the rate has always been determined by the exponentially decaying thermal output of the Earth. It is recognized that mantle convection and related crust-forming processes are driven by the Earth's radiogenic heat engine and model 1 uses this as the determinant of crustal growth (curve 1, Figure 1). Moorbath (1977, 1978) has argued that the abundance of mantle-type isotopic ratios in granites of all ages supports this model and Windley (1977) has found sufficient tectonic-petrogenetic evidence to concur. In this model, crustal accretion rates above subduction zones are, and always have been, numerically equal to the rates of growth: the continents are permanent.

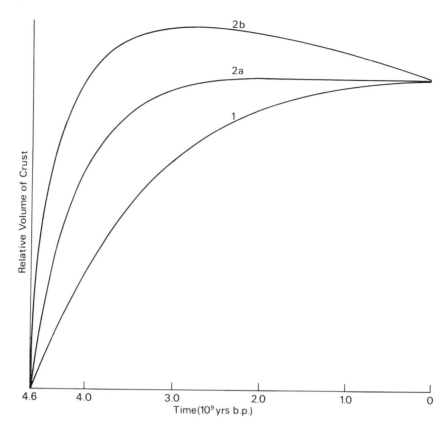

Figure 1 Schematic illustration of possible growth models for continental crustal volumes, adapted from Fyfe (1978). Curve 1 is the steady-state growth model and is exponential, curve 2a postulates the maintenance of an equilibrium volume after 2500 Ma ago, and curve 2b requires that continents are shrinking by erosion and subduction back into the mantle. For further details, see text

Model 2 suggests that most granitic crust was established rapidly, early in Earth history, and that it reached an equilibrium volume *ca* 2500 Ma ago (curve 2a, Figure 1). There are two possible explanations for this equilibrium volume: either (a) the continental crust is being recycled from within so that granite batholiths are furnished from the crust, or (b) the continents are eroded as fast as they are accreted from the mantle, the material removed returns to the mantle via subduction zones and is recycled as plate margin magmas. On the basis of sediment volumes in the geological record, Hargraves (1976) suggested that there was a world-encircling ocean in the Archaean but that continental emergence provided the opportunity for significant erosion and sedimentation to take place in subsequent times (cf. also Fryer *et al.,* 1979); this supports reason (b) above.

Fyfe (1978) has produced an interesting variant of model 2 (see curve 2b, Figure 1) by recognizing that, at the present day, continents are eroded at 2 km³

a^{-1}, four times as fast as accretion takes place. If a large proportion of the eroded sediments are recycled back into the mantle then the continent volumes may be shrinking.

To some extent, the evidence allowing a choice between these alternatives is equivocal but, on balance, the present writer favours model 1 for three reasons:

1. It is demonstrable that granites cannot easily be produced by crustal melting, thus removing pillar (a) from model 2. Figure 2 (cf. Brown 1979a) illustrates the actual temperature and granite-melting conditions for crust unperturbed by active magma injection from beneath. Two geotherms are drawn for lithosphere 100 km and 200 km thick assuming, in each case, that the base is isothermal and at basalt-melting temperatures (cf. Sclater and Francheteau, 1970). Average thermal gradients through the crust ($11-18°C$ km^{-1}) are inadequate for granites to melt within the upper 35 km. This does not invalidate any magma generation model that requires crustal mixing with mantle-derived liquids; rather, it shows that such liquids rising into fusible lower crust at perhaps $1000°C$, are capable of scavenging such a layer. Granites are, therefore, predominantly mantle-derived and, as noted earlier, the isotopic evidence for such an origin is almost unequivocal.

2. This focusses attention on the need to subduct ocean crust in order to maintain model 2. Here there is less certainty and while ocean crust layers 2—4 are undoubtedly consumed, the layer 1 sediments, of continental derivation and of much lower density, are often added back to continental margins as uplifted geosynclinal prisms into which later granites are sometimes intruded (see Miyashiro, 1973 and, specifically, for Japan: Oba, 1977, and for North America: Armstrong et al., 1977). The lack of sedimentary accretions on the Andean Pacific margin is not evidence of sediment subduction because the East Pacific Rise has received little sediment for the last 70 Ma (Ruiz Fuller, quoted in Zambrano and Urien, 1970). Only thin sequences, lacking obvious continental clastic material, were found in the DSDP Leg 35 cores adjacent to Peru and Chile. Gibbs (1967) has pointed out that the Amazon river is responsible for 20% of the net discharge of sediment into the World Ocean and so the Andes are eroding into the Atlantic. Recent estimates of Amazon sediment volumes by Meade et al. (1979) support this view. The case against sediment subduction and recycling as calc-alkaline magma cannot easily be proved. Yet, if the process is valid, then it is rather surprising that Andean magmas, generated in a subduction environment where the subducting ocean crust is depleted in continental erosion products, are so similar to those found elsewhere.

Other large volumes of marine sediment are found in zones of continental plate collision. For example, it is well known (De Jong, 1973) that marine sediments were uplifted, deformed and thrust over the closing continental forelands during the formation of the Alps and Himalayas. Whether or not this fate awaits the voluminous shelf sediments of the South Atlantic is a debate that cannot detract from the obvious conclusions: (a) that vast volumes of marine sediment are preserved in the geological record, and (b) that calc-alkaline magma types over modern subduction zones are independent of ocean sediment availability but

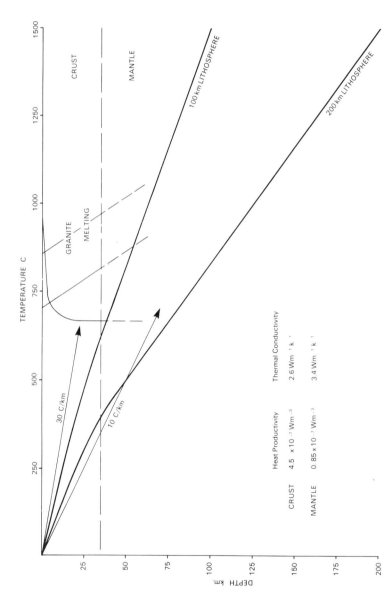

Figure 2 Lithospheric geotherms for 100 km and 200 km total thickness, calculated using 2-layer models each with an upper 35 km of radio-element-enriched crust (further details in Brown and Hennessy, 1978; Brown, 1979a). The wet granite melting curve, biotite and amphibole dehydration curves (in order of increasing temperature) are shown, together with appropriate geothermal gradients, for comparison

correlate with crustal thickness. At most, small amounts of sediment subduction and recycling are envisaged: this does not detract from the overall favourability of model 1 (Figure 1).

3. Finally, there is a sensible relationship between the modern continental accretion rate (ca 0.5 km^3 a^{-1}) through calc-alkaline magmatism and the total volume of the continents (ca 5 x 10^9 km^3) which is consistent with model 1. At the present accretion rate it would take 10^{10} years to 'grow' the continents but, because of the probable exponential decay of active geological processes, it is reasonable that the time-average accretion rate over the Earth's history has been twice that at present (cf. discussion by Brown, 1979a). In other words, accretion to the continents approximates to volumetric growth as illustrated by curve 1 in Figure 1.

It is suggested, for the following three reasons, that batholith emplacements are, and always have been, the major cause of progressive crustal growth throughout Earth history:

1. The difficulty of generating crustal melts in batholithic volumes.
2. The evidence that major volumes of continental erosion are not involved in subduction and recycling.
3. The fact that modern accretion rates seem to fit growth curve 1 (Figure 1).

Ancient batholiths: some variations and comparisons

Most modern granites are calc-alkaline and have the common diagnostic geochemical characteristics discussed earlier. However, not all ancient major crustal granites have these characteristics, some are not even calc-alkaline but are more alkaline. Figure 3 compares granites of various types and ages in terms of the range in their calcium/alkali ratios. All igneous rock series record a decrease in this ratio with silica content but the nature of this change distinguishes between different groups (cf. the alkali-lime index of Peacock, 1931). Four examples each of alkaline and calc-alkaline granite series are illustrated in Figure 3. These were selected as typical, but well-studied, examples from both Precambrian (solid lines) and Phanerozoic (dashed lines) terrains. Contemporary examples are the Sierra Nevada batholith (Bateman and Dodge, 1970) and the Skye Hebridean granites (Wager et al., 1965). The Skye granites do not reach batholith proportions, but they are included to indicate a geochemical similarity to more voluminous, ancient alkaline granites (see below).
    Concentrating first on calc-alkaline suites: older examples plotted in Figure 3 are the Ben Ghnema batholith of the Libyan Pan-African belt (Ghuma and Rogers, 1978) and the early Proterozoic tonalites of SW Finland (Arth et al., 1978). These are typical of gabbro-diorite-tonalite-trondhjemite suites that occur throughout geological time from the plutonic component of Archaean granite–greenstone belts (Tarney, 1976) up to the present day. There have been various

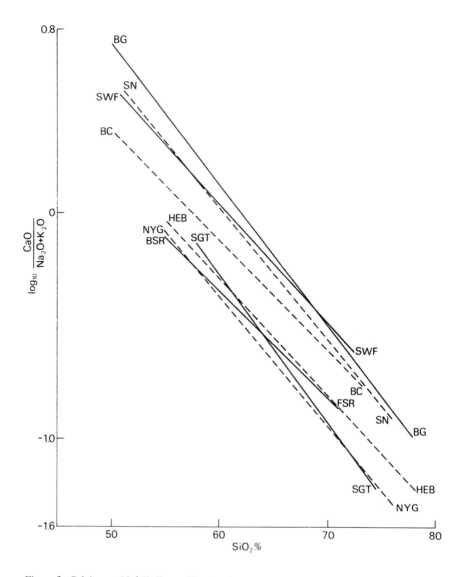

Figure 3 Calcium oxide/alkalis vs. silica plot for selected granitic suites showing the distinction between calc-alkaline and alkaline suites. Dashed lines represent Phanerozoic suites and solid lines represent granites from the Precambrian. BG = Ben Ghnema (Tibetsi, Libya); SWF = SW Finland tonalites; SN = Sierra Nevada; BC = British Caledonides; SGT = Saudi granite traverse (Arabian shield); FSR = Fennoscandian rapakivi granites; HEB = Hebridean Tertiary granites; NYG = Nigerian younger granites. Full source references are quoted in the text

explanations of Archaean calc-alkaline magmatism ranging from the generation of thin and static granitic rafts over mantle hot spots (Fyfe, 1978) to a highly mobile tectonic regime with subduction zones and small, thick continents (see explanation by Windley, 1977). In view (a) of the apparent high-pressure metamorphic assemblages from Archaean granulite terrains indicating a thick crust (Tarney and Windley, 1977), and (b) of the similarity of calc-alkaline Archaean greenstone magmas to those erupted behind modern destructive arcs, the most probable explanation involves a kind of plate tectonic mechanism. Particularly commendable are the models of Tarney *et al.* (1976) and Burke *et al.* (1976), which envisage that greenstone magmas were emplaced and erupted in extensional back-arc marginal basins over subduction zones. All the isotopic characteristics of these Archaean granite suites are also analogous to those of contemporary batholiths in supporting a predominantly mantle origin (Moorbath, 1977; Wilson *et al.*, 1978).

Calc-alkaline granites continued to appear along the margins of increasingly large, stable continental blocks during Proterozoic times (e.g. Barker *et al.*, 1976). In addition to the Proterozoic suites plotted in Figure 3, the Arabian Pan-African granites should be mentioned because their age, isotopic and spatial characteristics seem to indicate a multiple arc origin (Greenwood *et al.*, 1976; Gass, 1977). The consensus view from all geological ages is that calc-alkaline batholith magmas characterize subduction margins. Their origin is linked to parental magma generation in the upper mantle followed by contamination and/or fractionation during ascent through the continental crust.

Although these conclusions have widespread applicability, it is interesting that the earliest granites of theArabian shield (*ca* 950 Ma) are not calc-alkaline. These rocks have been studied in an E-W traverse across the shield by Gass (1977 and unpublished data) and their affinities (Figure 3) are strictly alkaline. Yet they form a major part of the Pan-African crystalline basement of Arabia. Likewise, the late Pan-African granites of the Sudan (Neary *et al.*, 1976) and Egypt (Greenberg, 1978) have alkaline affinities. Many other, older Proterozoic granites are also alkaline and are members of a voluminous gabbro-sodic anorthosite-syenite-rapakivi granite series for which the type area is Fennoscandia (Emslie, 1978). Many other examples are known from the mid-Proterozoic: for example, in New Mexico (Condie, 1978) and Colorado (Barker *et al.*, 1975). Generally, these suites have an intraplate geochemical imprint and are often associated with vertical, rather than lateral tectonics (Bridgewater *et al.*, 1974). Most of the major alkaline granite suites are of Proterozoic age: there are few volumetrically significant older or younger examples. More recent alkali granites are known to characterize either anorogenic sites, abortive or incipient rift systems (e.g. the Jurassic Nigerian younger granites, Macleod *et al.*, 1971; the Hebridean granites, Wager *et al.*, 1965) or continental collision zones where crustal melting is undoubtedly involved (Hamet and Allègre, 1976; Beckinsale, this volume). Emslie (1978) suggests, quite credibly, that the volumetrically significant suite of Proterozoic alkaline anorogenic–continental rifting granites was generated by melting from deep mantle sources, producing olivine tholeiite or alkali basalt liquids that fractionated and became contaminated at crustal levels. Their isotopic characteristics often, but not always, indicate a mixed crust–mantle origin with a higher crustal component than for calc-alkaline suites.

This discussion has indicated that granite types reflect tectonic environment, that most granites are calc-alkaline, due to shallow melting on continental margins, but that alkaline granites are important in Proterozoic intraplate locations. What was special about the Proterozoic tectonic regime? The change in tectonic patterns at the Archaean–Proterozoic boundary probably reflects the most significant development in crustal evolution. Following the highly mobile Archaean regime, the Proterozoic was dominated by broad, stable platforms and large continental landmasses that were resistant to fragmentation (Kroner, 1976 and palaeomagnetic evidence: Piper, 1976; see also discussion by Windley, 1977). Although geosynclinal basins and subduction zones developed around the Proterozoic landmasses, together with calc-alkaline magmatism (e.g. Hoffman, 1973), the landmasses themselves carry mobile belts where vertical tectonics, basin and swell features, predominate. There is little evidence for lateral movements across the African mobile belts, for example, and these were the sites of alkaline magmatism. Towards the edges of the large continental plates, alkali granites and mobile belts give way to more typical calc-alkaline suites and lateral deformational tectonics (Davies and Windley, 1976).

The difference between Archaean and Proterozoic tectonic patterns has been linked to the number of penetrative sites for mantle convection cells (e.g. Fyfe, 1978). Because of thermal decay, the number of cells would have diminished from Archaean to Proterozoic times and so zones of magma penetration would have become more widely spaced. However, there is a scale problem because Archaean penetrative domes are on a scale of $10-10^2$ km (Wilson *et al.*, 1978) whereas the Proterozoic landmasses were 10^3-10^4 km across (Piper, 1976). The mantle must have been convecting beneath these landmasses and the true scale of convection cells may be reflected by the more closely-spaced mobile belts. In that case, it is surprising that fragmentation did not occur and the continental lithosphere must have been inefficiently coupled to the mantle convective drive. This raises the question of how long this state of affairs persisted. There is good evidence that the continents have been mobile for the last 200 Ma, but before that a major period of Phanerozoic continental aggregation into Pangaea is postulated (Smith and Hallam, 1970). Part of the aggregation process was the suturing of the Iapetus (Caledonian) ocean across the northern continents *ca* 450 Ma ago (Dewey, 1969; Phillips *et al.*, 1976). A major suite of *ca* 400 Ma British granites, generated just after this closing event, is plotted in Figure 3 (cf. Stephens and Halliday, this volume; Brown, 1979b). In many ways these granites reflect the space-time trends of modern plate margin batholiths, although they differ in some respects. In particular, they post-date the closure event, which itself was not diastrophic on the scale of the modern Himalayas, and the granites are of a transitional geochemical type but with a calc-alkaline flavour (Figure 3). Apparently, the Caledonian ocean closed rather passively, perhaps again reflecting inefficient coupling to mantle convection, and it was necessary for crustal thickening to occur over the former active spreading zone before granite melts could be generated (Brown, 1979b). Read (1961) recognized these intrusions as 'permitted' and it may be that mildly tensional tectonic conditions are reflected in the geochemical affinities of this early Phanerozoic magma suite.

114

Summary of conclusions

1. There is good evidence that primary granite batholiths have been formed and that resulting progressive, long-term continental growth has occurred throughout Earth history. However, the process has decreased in intensity as heat productivity inside the Earth has decayed exponentially.
2. Most volumetrically significant 'granite' batholiths are generated predominantly by mantle fusion. Two magma types are recognized:
   a) calc-alkaline suites generated in the shallow mantle wedge above lateral compressive zones of ocean crust destruction;
   b) alkaline suites from anorogenic, tensional swell environments whose magma source lies within deeper mantle zones.
3. Calc-alkaline batholiths have predominated for most of Earth history, particularly during the Archaean period of major crustal stabilization, but alkaline suites became important products of intraplate magmatism during mid-Proterozoic times (*ca* 1500–2000 Ma ago). This may reflect relatively inefficient coupling between lithosphere and asthenosphere during the Proterozoic.
4. The geochemistry of late Proterozoic Pan-African and early Phanerozoic Caledonian granites is transitional but with a bias towards subduction-related calc-alkaline magmatism. Phanerozoic granites have become more abundantly calc-alkaline as continental fragmentation and increased mobility developed. This may be the final stage in the long-term evolution of continental magmatism over the last 2000 Ma, during which time frictional coupling between continental tectosphere and mantle convection has increased.

AN EMPLACEMENT MECHANISM FOR POST-TECTONIC GRANITES AND ITS IMPLICATIONS FOR THEIR GEOCHEMICAL FEATURES

M.T. HOLDER
Institute of Geological Sciences, St. Just,
30 Pennsylvania Road, Exeter EX4 6BX, UK

Introduction

Concentric zoning is a prominent feature of post-tectonic granitoid plutons and its origin has commonly been ascribed to processes of differentiation and/or contamination. It is the purpose of this contribution, however, to examine what effect the intrusion mechanism of post-tectonic plutons has in the production of such zoning. To this end, the Ardara pluton, which has become one of the classic examples of a zoned post-tectonic pluton (Akaad, 1956), has been studied from a structural point of view.

The Ardara pluton lies within the cluster of Caledonian granites in NW Donegal. The outcrop is roughly circular with a 'tail'-like apophysis in the SW (Figure 1), and is almost entirely surrounded by Dalradian pelites and semi-pelites of the Cleengort group. The only exception to this is where the 'tail' is in contact with an appinite body: the Meenalargan complex.

Three distinct compositional units make up the pluton. Outermost is a leuco-cratic, coarse-grained and strongly megacrystic quartz-diorite (G_I) which is in sharp and smooth contact everywhere with the surrounding pelites and semi-pelites. This unit has a ring-like outcrop which is marginal to the pluton, except in the area of the 'tail'. Within this ring a coarse-grained, megacrystic granodiorite (G_{II}) outcrops in a broad horseshoe shape, open to the south and occupying the 'tail'-like apophysis. Its contact with the surrounding quartz-diorite is always sharp but is sometimes interdigitated, while its contact with the Meenalargan complex in the 'tail' is sharp and irregular, occasionally developing agmatites. Lying inside the outcrop of this unit, in the centre of the pluton, is a coarse equigranular granodiorite (G_{III}). It, too, has a sharp contact with the quartz-diorite (G_I) in the south but has gradational contacts with G_{II}.

Tectonism associated with the pluton

In the NW and E of the pluton the pelites and semi-pelites of the Cleengort group have been strongly deformed around the granite margin. Those regional folds and cleavages, which initially were parallel to the contact, have been strongly flattened, while new folds and cleavages have been formed elsewhere to produce an intense penetrative cleavage wrapping around the intrusion. In the

116

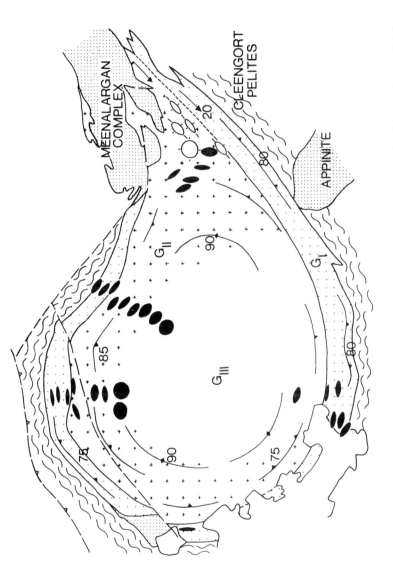

MEENALARGAN COMPLEX

CLEENGORT PELITES

APPINITE

G_I

G_{II}

G_{III}

20

80

90

85

90

75

75

80

Figure 1 The Ardara pluton: numbers refer to the dip of internal foliations and ellipses represent the finite strain recorded by xenoliths (outcrop after Akaad, 1956, reproduced with permission of the Editors, Pitcher and Berger). Filled ellipses record deformation associated with intrusion of the Ardara pluton; open ellipses deformation associated with intrusion of Main Donegal Granite

south, however, later deformation linked to the intrusion of the neighbouring Main Donegal Granite has destroyed this alignment and replaced it with a strong NE-SW LS fabric (Flinn, 1965). There is a concentric planar gneissic schistosity within the pluton, parallel to the cleavage in the country rocks. It is defined by the planar alignment of plagioclase megacrysts, elongate quartz aggregates and biotite flakes. Undulose extinction in quartz and bending and granulation of plagioclase indicates that this schistosity is tectonic in origin, while the lack of mineral alignments within the plane of the schistosity shows it to be an 'S' fabric (Flinn, 1965). At the pluton margin it is strongly developed, steep and inwardly inclined, steepening to the vertical and becoming weaker towards the centre of the granite mass; it finally disappears to leave a large unfoliated central core. This fabric is not limited to any compositional zones within the pluton and is seen to cross-cut internal contacts where these are not parallel to the outer granite contact.

Xenoliths within the granite, of locally-derived basic igneous rocks and psammites, have been deformed into flattened biaxial ellipsoids. Since many of the xenoliths show no effects of competence contrast with their granite matrix, it has been possible to use them as strain markers for the flattening deformation which has caused the planar schistosity to develop within both the granite and the marginal country rocks.

The finite strain was determined using the geometric mean method at each locality by measurement of the longest and the shortest axes of each xenolith in a horizontal plane, parallel/subparallel to the XZ/YZ section of the finite strain ellipsoid. This method was considered adequate for the majority of localities where the XY plane of the xenoliths was parallel to the schistosity, but in the few cases where this was not the case the Rf/ϕ method of Dunnet (1969) was employed. Measurements within the XY plane of the xenoliths were limited by the largely two-dimensional exposure, but when all measurements from the XY plane are considered together, no alignments of X axes are apparent despite reports (King, 1966) that vertical lineations do exist in the magnetic fabric of the granite. This being so, the finite strain is of pure flattening type and measurements in any plane perpendicular to the XY plane will define the strain ellipsoid.

Populations of xenoliths of between 15 and 43 were used for finite strain determination, dependent on the abundance of suitable xenoliths at each locality. Inaccuracies caused by low numbers of xenoliths were calculated by taking similar numbers of measurements, at random, from a larger population and determining the strain from them in comparison to that calculated from the entire population at that locality. By this method the maximum error caused by low numbers of xenoliths was found to be 15%. Errors caused by the initial non-spherical shape of xenoliths were assessed by comparing the strain at each locality determined from basic xenoliths (which are likely to have had an initial block-like shape), with that determined from psammite xenoliths (which have a more slab-like initial shape). Despite the higher initial axial ratio of the psammite xenoliths, no consistent variation in finite strain was found between them and the basic xenoliths, indicating that initial shape does not control the measurement of finite strain to any significant degree.

Figure 1 gives the results of the determination of the finite strain for localities within the pluton and shows that, in common with the intensity of schistosity development, the strain decreases inwards from the margins of the pluton, becoming zero in the unfoliated central zone.

Chronology of tectonothermal events in the pluton

Within the pluton, the lack of chronological markers inhibits the study of the development of this gneissic schistosity. In the metamorphic aureole, however, the metamorphism itself provides a time marker. The development of the concentric schistosity can best be studied, therefore, by an investigation of metamorphic textures. Three areas have been studied in the NW aureole: one in the andalusite zone, another in the sillimanite zone and the last within the sillimanite-free zone adjacent to the granite contact.

ANDALUSITE ZONE

The andalusite zone is characterized by an assemblage of andalusite, staurolite, garnet, fibrolite, biotite and white mica. Andalusite, staurolite and garnet all possess a 'core and rim' structure. From the relationship between staurolite and andalusite it is clear that the cores of these minerals developed during the same event and that they grew over a strong penetrative fabric. Inclusions within the rims are deformed around the cores and show that the fabric overgrown by the cores was rejuvenated prior to the second metamorphic event. Post-dating this prograde metamorphism, a minor retrogression caused staurolite to be partly retrogressed, garnet to be replaced by biotite, and andalusite to be partly replaced by biotite and white mica. A further rejuvenation of this fabric caused the rims of andalusite, garnet and staurolite to be deformed. This was followed by the third metamorphism causing fibrolite to form at the expense of biotite and white mica which surround the andalusite. The end of contact metamorphism is indicated by a second retrogression in which andalusite is again partially replaced by biotite and white mica, fibrolite by white mica, and biotite partially retrogressed to chlorite. This sequence is illustrated in Table 1.

SILLIMANITE ZONE

At the boundary of the sillimanite zone the deformation of the country rocks increases rapidly in intensity, the boundary forming the 'hiatus' of Akaad (1956). This sudden increase seems to have been caused by (a): the position of a major syncline of regional age, to the NE of the pluton, which protects the outer aureole from much of the deformation; together with (b): faults in the N and NW, active during intrusion and behaving in a similar manner. It may also reflect a change in the ductility of the country rocks due to the temperature increase in the inner aureole.

Within the sillimanite zone the textural relationships are somewhat simpler than in the andalusite zone. The characteristic minerals occurring in this zone are: fibrolite, white mica, biotite, garnet and plagioclase. Andalusite and staurolite are

Table 1

	Andalusite zone	Sillimanite zone	Sillimanite-free zone
$^{RMP}G_{III}$	Replacement of: fibrolite by w. *mica* biotite by *chlorite* andalusite by w. *mica* and *biotite*	Replacement of: fibrolite by w. *mica* biotite by *chlorite*	Replacement of: sillimatite and alkali feldspar by w. *mica* and *quartz*
$^{MP}G_{III}$	*Fibrolite* growth	*Fibrolite* growth	Growth of: *garnet, plagioclase, (alkali feldspar, sillimanite?)*
$^{S}G_{III}$	Deformation	Deformation	Deformation
$^{RMP}G_{II}$	Replacement of: staurolite, garnet by *biotite* andalusite by w. *mica* and *biotite*	Replacement of: andalusite by w. *mica* and *biotite* garnet by *biotite* and growth of w. *mica*	Replacement of: andalusite by w. *mica* and *biotite* garnet by *biotite* and *opaques*
$^{MP}G_{II}$	Growth of: *garnet, andalusite, staurolite*	Growth of: *garnet, andalusite*	Growth of: *garnet, andalusite*
$^{S}G_{II}$	Deformation	Deformation	?
$^{MP}G_{I}$	Growth of: *garnet, andalusite, biotite, staurolite*	?	
$^{S}G_{I}$	Deformation?		
	Regional metamorphism and deformation	Regional metamorphism and deformation	?

absent but are represented as pseudomorphs of biotite and white mica. Again, the rock possesses a strong penetrative fabric and a crenulation deforming these pseudomorphs which include trails of opaque minerals marking an earlier oblique fabric. Garnet is strongly broken and partially replaced by biotite in more pelitic horizons, or by plagioclase of An_{35-40} composition where biotite is absent. Fibrolite is ubiquitous, growing at the expense of the biotite which defines the main fabrics, and also replacing garnet. It forms clots and ramifying veins which are completely undeformed. The close of metamorphism is marked by a retrogression of fibrolite to white mica and a partial retrogression of biotite to chlorite. A summary of this metamorphic sequence and its relationship to that seen in the andalusite zone is given in Table 1.

Immediately adjacent to the pluton, sillimanite disappears and the assemblage becomes biotite, white mica, garnet, plagioclase and quartz. Fabrics in this zone are weak and pre-date ragged porphyroblasts of plagioclase (An_{37}) and white mica. The latter display vermicular intergrowths with quartz, possibly from the replacement of sillimanite and alkali feldspar. In some rocks euhedral white micas seem to have grown at the expense of biotite, possibly pseudomorphing prismatic sillimanite, while other rocks display advanced anatexis. Garnets within this zone remain as broken fragments, heavily replaced by biotite. The lack of either sillimanite or alkali feldspar in these rocks, which in places have commenced melting, seems to be due to the efficiency of the closing phase of retrogression.

The analysis of metamorphic textures seen in the aureole of the Ardara pluton is summarized in Table 1. From this summary it is clear that at least two separate coaxial deformations occurred within the country rocks during the intrusion of the Ardara pluton. Moreover, since these deformations only affect the country rocks marginal to the pluton and post-date the regional deformation, it is evident that they were caused by the intrusion of the pluton. Similarly, three distinct prograde metamorphisms occurred, separated by periods of deformation. The time gap between these metamorphisms was certainly large as indicated by the retrogression between the second and third prograde metamorphisms. The inescapable conclusion is that the intrusion of the Ardara pluton was accomplished in three separate events, during each of which the country rocks were forcefully displaced by the incoming magma.

Interpretation of strain results from the pluton – Ramsay's 'balloon'

The tectonothermal events have also been recorded in the deformation of xenoliths within the pluton itself. These xenoliths were deformed after being trapped by consolidating granite. Thus, assuming that the xenoliths were in free fall in the magma and that the magma consolidated first at the edge of the pluton, xenoliths at the margins of the pluton will have suffered a longer period of deformation, because they were trapped earlier than those nearer the centre. Ramsay (in press), in his treatment of the Chinamora batholith, used these assumptions to construct a model for granite emplacement by inflation, or 'ballooning', of the intruding pluton – an idea originally proposed by Cloos (1925) for the Riesengebirge granite. Ramsay approximates the growing pluton to an expanding sphere since the magma pressure driving the inflation and country rock deformation is hydrostatic. Expansion of the sphere will cause any line on its surface to be extended by equal amounts, while lines perpendicular to the surface will be shortened, i.e. pure flattening deformation. The amount of deformation of any part of the sphere is directly proportional to the increase in the size of the sphere.

In order to illustrate this model we will consider a pluton at two stages during its growth (Figure 2). At a time (t), early in the intrusion of the pluton, the total radius of the pluton is (a). An undeformed xenolith (X:Y:Z = 1) is trapped close to the margin of the pluton at radius (d) as the freezing surface moves inward

$$X.Y.Z = 1 \;\therefore\; r.r.1/r^2 = 1$$
$$\frac{X}{Z} = \frac{Y}{Z} = r^3 = \text{'strain'}$$
$$\frac{D}{d} = \frac{r}{1} \;\therefore\; (D/d)^3 = \text{'strain'} \;\therefore\; d = D/\text{strain}^{-3}$$

Volume of granite at time $t = f(a)^3$ at $T = f(A)^3$

Volume of solid granite at $t = f(a^3 - d^3)$

$$= f(A^3 - D^3)$$

$$\therefore f(a)^3 = f(A^3 - D^3 + d^3) \;\therefore\; a = (A^3 - D^3 + d^3)^{-3}$$

Figure 2 Ramsay's balloon model: the geometric relationships between finite strain in xenoliths and the size of a pluton at two stages during its growth

from the edge of the pluton. At a later time (T) the total radius of the pluton has expanded to (A). The same xenolith has also been moved, due to the inflation, to a new radius (D) and this has caused both it and the enclosing granite to suffer a pure flattening deformation. Meanwhile, the granite has continued to cool causing the freezing surface to move inwards and trap further xenoliths. The axial length of the xenolith after this amount of deformation at time (T) thus becomes $r;r;1/r^2$, since the products of the axes must remain equal to unity if there is no volume loss during deformation. Since the angle subtended at the centre of the pluton by the X and Y axes of the xenolith, lying in the plane of the surface of the sphere, will remain the same both before and after deformation, the change in their length $(r/1)$ can be written as (D/d). The finite strain (X/Z) or (Y/Z), can therefore be written as $(r/(1/r^2))$, i.e. (r^3).

Since $r = D/d$ then $r^3 = (D/d)^3$
Therefore, finite strain $= (D/d)^3$
Rearranging this, $d = D/\text{strain}^{-3}$.

Since, at the end of intrusion, (D) is the distance of the xenolith from the centre of the pluton at the present time, and the finite strain represented by the deformed xenoliths can be measured, it is possible to calculate the radius of the freezing surface at the time the xenolith was trapped. Using xenoliths across the pluton we can therefore trace the variation of the radius of the freezing surface throughout the period of intrusion.

Because our pluton model is spherical, its volume at time (t) can be represented as V_t,

where $V_t = f(a)^3$
at time (T), $V_T = f(A)^3$.

The volume of granite which has consolidated at time (t) is given by the equation,

$V_t' = f(a)^3 - f(d)^3$

This volume must be the same as that lying outside the radius of our xenolith at time (T), since the xenolith marks the position of the freezing surface from time (t).

This volume is $V_T' = f(A)^3 - f(D)^3$
Since $V_t' = V_T'$
Then $f(a)^3 - f(d)^3 = f(A)^3 - f(D)^3$
Rearranging this we obtain, $a = (A^3 - D^3 + d^3)^{-3}$

Knowing both (A) and (D) from direct measurement and (d) by calculation, it is now possible to plot the variations in both the radius of the freezing surface and the total radius of the pluton throughout the entire period of intrusion.

The intrusion history of the pluton

Applying Ramsay's model, as described above, to the Ardara pluton, we obtain three maxima in the radius of the freezing surface (Figure 3), indicating that the growth of the pluton has not been uniform. The increases in the freezing surface radius cannot be explained by variations in the cooling rate since they correspond with increases in the total size of the pluton and must therefore be related to increases in the volume of the pluton by periodic influxes of magma. The position of the influx of these magma pulses in Figure 3 is coincident with the position of the internal contacts between the granitic compositions of the pluton. This, however, does not necessarily mean that each pulse was composed of only one compositional type. If the quartz-diorite is taken as an example, we can see from Figure 3 that xenoliths marking the end of the first pulse lie on the G_I/G_{II} contact. At this point all the quartz-diorite now present in the pluton had consolidated and any further crystallization must have created granodiorite. At this time, however, a large volume of magma still existed in the centre of the pluton as the residue of the first pulse. If this residue was granodiorite it must imply that both the quartz-diorite and granodiorite existed as melts side-by-side in the first pulse without mixing, since the contact between them is extremely sharp and because they do not share xenolith types. An alternative possibility is that the magma residue was of quartz-diorite composition, but that the influx of the second pulse, of granodiorite composition, caused this residue to be swept upwards, leaving only the consolidated parts of the pluton behind.

This view is supported by the small volume of quartz-diorite represented in

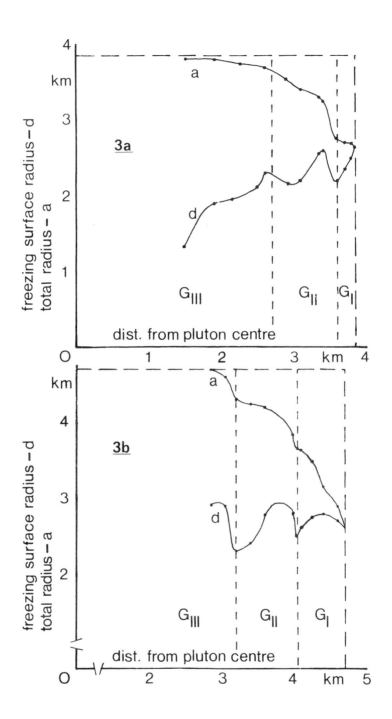

Figure 3 (a) and (b). Data from two traverses in the Ardara pluton showing the variation in the freezing surface radius and total size during intrusion (stages of intrusion becoming younger towards the left of the graphs)

the pluton (calculated as a spherical shell) of 50 km³, as compared with 85 km³ of G_{II} excluding the volume within the 'tail', and 105 km³ of G_{III}. Since, at the end of the intrusion of the first pulse, the volume of the pluton was 94 km³, it would seem likely that all three pulses were of similar volume but that almost half of the first pulse was lost to a higher level. No magma seems to have been lost during the intrusion of G_{III} into the still molten centre of G_{II} and mixing and sharing of xenoliths has taken place along the contact between them.

A further peculiarity of the first pulse is displayed in the freezing surface variation curves in Figure 3(a). These show only a decrease in the radius of the freezing surface during the intrusion of the first magma pulse, while Figure 3(b) shows both an increase and a decrease of this feature. This implies that part of the record of the freezing surface variation is missing from the area of the pluton used to construct Figure 3(a). In order to omit this part of the record, xenoliths in this area must have remained undeformed while the pluton was inflating, i.e. consolidation of the granite commenced later in this area than in that used for Figure 3(b). Moreover, the absence of any chilling of the quartz-diorite against the country rocks argues that this phase existed as an unconsolidated magma everywhere in the pluton for some time before cooling below its solidus, to produce coarse megacrystic textures right up to the country rock contact. This means that although Figure 3(b) displays more of the history of the quartz-diorite than Figure 3(a), it must also omit some of the earlier stages of inflation of the G_I phase achieved before consolidation began.

If this is true for G_I it is also likely that a similar period of inflation without consolidation affected both the G_{II} and G_{III} pulses, intruded as they were into rocks which must have been hotter than those encountered by G_I. The G_{II} phase shows evidence of having remained unconsolidated for a period after intrusion, as it was sufficiently mobile, after breaking through the G_I shell and forming the 'tail' apophysis, to stope large numbers of xenoliths from the Meenalargan complex and then to distribute them around its outcrop. G_{III} shows a similar effect in its gradational contact with G_{II} and acquisition of xenoliths from it, which suggests that hybridization of the magmas occurred along their contact.

This delay in consolidation of the intrusion causes the curves in Figure 3 to be discontinuous since a certain part of the inflation of each pulse will not have been recorded by the deformation of xenoliths. Figure 4 shows the reinterpretation of the freezing surface and total radius variation data to accommodate this feature of the intrusion process.

BRITTLE DEFORMATION IN THE INTRUSION PROCESS

As described so far, the intrusion mechanism of the Ardara pluton has been achieved by plastic deformation, but in the region of the 'tail' the G_I shell has been ruptured and G_{II} has invaded the Meenalargan complex. This rupturing is responsible for the incorporation of basic xenoliths into the G_{II} phase and thus into the outer zone of G_I. Such a brittle deformation is not accounted for by Ramsay's model and any increase in the pluton size accomplished by this method will not be recorded by the deformation of xenoliths. The effect of this fracturing of the G_I shell without internal deformation will have been to increase the radius of all

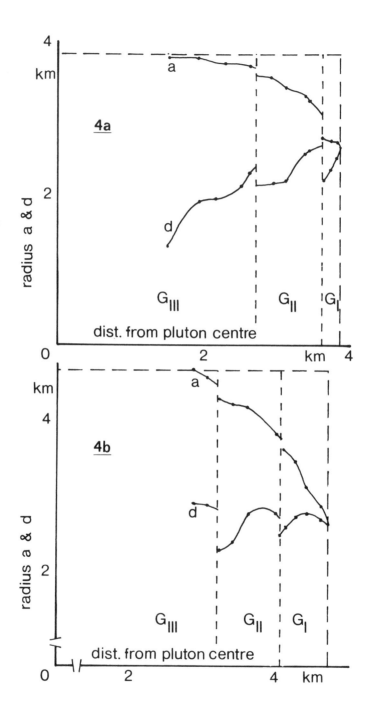

Figure 4 (a) and (b). Freezing surface radius and total size curves redrawn to allow for delayed consolidation of the inflating pulses. Note that the data used are the same as for Figure 3

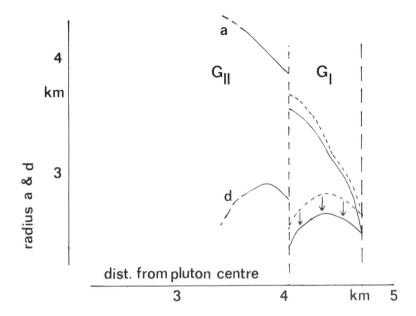

Figure 5 Enlarged plot of the G_I pulse from Figure 4(b). Dotted line shows the position of curves in Figure 4(b); solid line shows the position recalculated to allow for expansion during the fracture of the G_I shell by the intrusion of G_{II}

points within the quartz-diorite without increasing the strain. Since strain is equal to $(D/d)^3$ the increase in (D) will cause the calculated values of (d) and hence (a) to be too large. In order to correct for this effect the value of (D) must be reduced to remove the expansion due to opening of the fracture in G_I. At the end of intrusion the radius of the pluton is (A) and its circumference f(A). The circumference of G_I is, however, f(A) (360−x)/360 since there is a gap in its outcrop caused by rupturing which subtends an angle of x at the centre of the pluton, in this case 20 degrees. The value of (D) for a xenolith within G_I, therefore, must be corrected to D(360−x)/360 to allow for the opening of the 'tail'. Corrections can therefore be made for all the xenoliths measured within G_I, modifying the freezing surface and total pluton radius curves as shown in Figure 5, which indicates that this brittle deformation mechanism has had a small but significant effect on the growth of the pluton, especially as it allowed G_{II} and G_{III} to acquire their xenoliths.

Summary

From the preceding results there can be no doubt that the Ardara pluton has intruded by deforming its country rocks and that it has created at least 72% of its present volume by this process. Moreover, its intrusion was accomplished in three distinct tectonic and thermal events caused by the injection of three pulses of magma, each of which consisted of approximately 95–100 km³ of magma. Each of these pulses was injected at temperatures sufficiently in excess of the solidus to prevent consolidation until a certain degree of inflation had taken place. Despite this situation, little mixing between the magma pulses seems to have taken place.

Conditions within the intruding magma were thus ideal for crystal settling to take place and to have lead to large-scale differentiation. Shaw (1965) gives the viscosity limits for water saturated granitic melts as 10^{5-8} poise and states that for granites larger than 1 km in diameter, crystal settling rates are large. However, no differentiation is apparent within any of the three pulses at Ardara. Current activity in the cooling magma seems to offer the most likely explanation for this lack of differentiation. Bartlett (1969) claims that convection occurs in plutons of very small size even if the viscosity is 10^8 poise. Certainly, currents must have been active within Ardara to distribute xenoliths of the Meenalargan complex so widely throughout the G_{II} phase. Whatever mechanism is responsible, no differentiation of magma pulses has taken place despite the liquid nature of the intruding magma. The compositional zoning of the pluton has therefore been caused by the repeated intrusion of individual pulses of separate composition, interaction of which was limited.

Where the magma intruded the ductile Cleengort pelites the envelope has been deformed by plastic means, but wherever the country rocks are most competent, as in the Meenalargan complex, brittle mechanisms have been imposed and stoping has occurred. This fact is supported by the lack of pelitic xenoliths in the granite. It seems, therefore, that the relative ductility of the country rocks is the controlling factor which determines whether the dominant intrusion mechanism is inflation or stoping. It is only by the former process, however, that any record of the episodic nature of intrusion is retained by the granite.

ACKNOWLEDGMENTS

I should like to acknowledge Professor J.G. Ramsay whose previous work stimulated this study.

128

ISOTOPIC ANALYSIS OF TRACE SULPHUR FROM SOME S- AND I-TYPE GRANITES: HEREDITY OR ENVIRONMENT?

M.L. COLEMAN
Institute of Geological Sciences, 64-78 Gray's Inn Road,
London WC1X 8NG, UK

Introduction

Detailed petrological, geochemical and geochronological studies on rocks of the New England batholith of eastern Australia have resulted in a division into five suites, each identified by its own characteristic mineral assemblage and field relationships (for further background and bibliography see O'Neil *et al.*, 1977). The suites have been named after their typical locations: Bundarra, Hillgrove, Moonbi, Uralla and Stanthorpe. With the exception of the Stanthorpe group, each suite can be described as having originated from either a sedimentary or igneous source: the S- and I-types of Chappell and White (1974). The Stanthorpe suite is composed of late-stage leucocratic granites with possible I-type affinities.

Apart from differences in their mineralogy and chemistry, the two types of granite are also clearly distinguishable by their oxygen-isotope compositions (O'Neil *et al.*, 1977). In general, the study of the isotopic composition of sulphur in sulphide minerals associated with granites may give ambiguous information on the origin of the host rocks (Coleman, 1977), and this has proved also to be the case for the New England batholith (Herbert and Smith, 1978).

This paper is a preliminary report describing work done in collaboration with Dr. Stirling E. Shaw of Macquarie University, which involved the isotopic analysis of *trace* sulphur in the same rocks examined by O'Neil *et al.* (1977). Thirty-two samples were carefully taken, avoiding contact zones or areas where hydrothermal alteration may have occurred. As well as being categorized as either S- or I-type, the mineralogy of the plutons independently differentiates them into oxidized or reduced assemblages from the presence of ilmenite, as opposed to magnetite, and the Fe_2O_3/FeO ratio in the biotite. The association of S-type with ilmenite and I-type with magnetite (Ishihara, 1977) does not apply to these suites. The type and oxidation state of the five suites is shown in Table 1.

Sulphur-isotope analysis

The sulphur was extracted from some of the samples as sulphur dioxide by direct oxidation (Blatt and Brown, 1974; Coleman and Moore, 1978). However, in many cases the sulphur content was too low for this method so reduction to hydrogen sulphide and precipitation as silver sulphide was used (Thode *et al.*, 1961); the

Table 1 The five suites of the New England batholith of eastern Australia

Suite	S or I	Oxidized or reduced
Bundarra	S-type	Reduced
Hillgrove	S-type	Reduced
Moonbi	I-type	Oxidized
Uralla	I-type	Reduced
Stanthorpe	(I-type?)	Oxidized

sulphide was then oxidized to sulphur dioxide (Robinson and Kusakabe, 1975). Both methods were used on some samples and similar isotopic results were obtained regardless of the technique employed. It was also possible to estimate the sulphur content of the samples by measuring the yield from the extractions. The sulphur dioxide was analysed on a Micromass 602-C mass spectrometer, modified by the incorporation of a heated inlet system.

The sulphur content was generally low and ranged from less than 2 ppm to 2300 ppm. The sulphur-isotope compositions, expressed in the normal way as $\delta^{34}S$ with respect to Cañon Diablo meteorite, all fall in the range $-10.5‰$ to $+6.0‰$ In detail, four points can be made:

1. The reduced rocks generally have negative $\delta^{34}S$ values while the oxidized ones have positive values.
2. The sulphur-isotope compositions of the I-type granites fall into a narrow range, $-3.6‰$ to $+4.2‰$.
3. In contrast, the S-type granites range from $-10.5‰$ to $-5.7‰$ (with three exceptions, all of which contain less than 15 ppm sulphur).
4. The Stanthorpe leucogranite samples have sulphur of similar isotopic composition to that of the I-type granites, but they give a negative correlation with oxidation state.

Discussion

In order to interpret the data sensibly, some background of sulphur-isotope geochemistry is necessary. The sulphur-isotope values for meteorite troilite form a tight group around $0‰$; the sulphur in basic igneous rocks and volcanics also gives values near zero. The fractionation of sulphur-isotopes in sedimentary regimes leads to a very much wider range in values; down to $-50‰$ for sulphides in pelitic rocks and as high as $+35‰$ for sulphate in evaporites. This distribution is easily explained by reference to experimental work. Equilibrium fractionation between oxidized and reduced species gives components enriched and depleted in ^{34}S, respectively (positive or negative $\delta^{34}S$). In addition, the magnitude of the

130

fractionation is very small at high temperatures and increases greatly at lower ones (for a fuller treatment of this topic and bibliography, see Coleman, 1977).

Following from this, it is possible to postulate various models for producing the sulphur-isotope composition in granites and to test them on the data from this batholith. There are three types of process which may contribute to a significant extent in producing the final sulphur-isotope composition of the rock: they may be considered in terms of the influences of heredity and of environment.

1. Inheritance of isotopic composition from the parental material would give values near zero for magmatic or volcanoclastic material and values displaced from zero by the incorporation of sulphur from sediments: negative from reduced sediments and positive from evaporites. In practice, negative values would be more likely, given the common sorts of sediment usually available.
2. Fractionation during petrogenesis will only produce a relatively small effect, because of the high temperatures involved. However, the chemical environment may cause loss of vapour phase constituents. Under reducing conditions, expulsion of isotopically-light hydrogen sulphide will leave the residual sulphur relatively heavy. Conversely, in an oxidizing environment loss of sulphur dioxide will leave the resulting rock isotopically light. With even more oxidation sulphur dioxide may not be lost but could be fixed as sulphate, producing a value nearer to that of the original composition (Schneider, 1970). Thus, the extent of fractionation will depend mainly on the relative proportions of the fugitive and residual components and this, in turn, may vary with the redox of the environment.
3. Contamination of the melt with sedimentary material will give the same sort of values mentioned in (1). In this case, however, the oxidation state of the rock will be controlled by the chemistry of the sediments in proportion to the amount present.

To determine which of these processes is relevant to each of the suites of rocks, the isotope data are plotted against the oxidation state, measured as Fe_2O_3/FeO ratio in biotite (Figure 1). For all the suites except Stanthorpe, where individual points are plotted, the area is defined by one standard deviation of the grouped data along each axis. The I-type granites of the Moonbi and Uralla suites show relatively little spread of $\delta^{34}S$ in comparison with the range of oxidation state, and also show positive and negative values for the oxidized and reduced suites, respectively. This suggests direct inheritance of oxidized and reduced igneous sulphur-isotope ratios, with little subsequent modification. This is in agreement with their formation by an extensive amount of partial melting of the parental material; evidence for this is the presence of considerable volumes of intermediate rocks and xenoliths of similar composition (O'Neil *et al.,* 1977). The Hillgrove S-type suite shows a similar narrow spread of $\delta^{34}S$ but gives values more characteristic of sulphide in sediments, and hence may have been derived directly from parental sedimentary material. However, in the case of the Bundarra S-type rocks, the spread of $\delta^{34}S$ is quite large and, despite being slightly more reduced than those of the Hillgrove suite, they give a less negative average $\delta^{34}S$. The

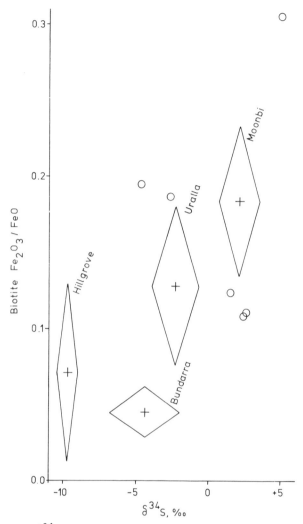

Figure 1 Variation of $\delta^{34}S$ with Fe_2O_3/FeO. The grouped data for each suite are defined in terms of plus and minus one standard deviation in both axes. The individual points are samples of the Stanthorpe Suite

lower sulphur content of these rocks (average, 9 ppm compared with 660 ppm for Hillgrove) is consistent with the argument that these rocks were sufficiently reduced to have lost isotopically light hydrogen sulphide thereby leaving the residual material with a more positive $\delta^{34}S$. In this case the inherited composition, which confirms the petrological results of Flood and Shaw (1975), has been modified by fractionation during petrogenesis.

The Stanthorpe leucogranites, too, show evidence of modification of the isotope ratios after inheritance. The $\delta^{34}S$ values alone suggest an I-type granite but the relationship between Fe_2O_3/FeO and $\delta^{34}S$ is informative. Starting with the sample having the lowest Fe_2O_3/FeO ratio, increase in oxidation state correlates negatively with $\delta^{34}S$, representing loss of greater amounts of isotopically positive sulphur dioxide. The sample with the highest Fe_2O_3/FeO ratio has probably retained oxidized sulphur as sulphate, and has a similar $\delta^{34}S$ to that of the least oxidized (and probably least modified) sample. The relationship between $\delta^{34}S$ and Fe_2O_3/FeO is probably due to the presence of a large amount of water in the oxidized I-type melt, which caused varying amounts of further oxidation during petrogenesis. This is supported by the very heavy oxygen-isotope composition shown by the most oxidized sample which could have lost light oxygen as water vapour (O'Neil et al., 1977). This same mechanism may account for the positive sulphur-isotope values in the magnetite series granitoids described by Sasaki and Ishihara (1979) who could not readily explain these data.

It can be seen that the oxidation state of the parental material is reflected in its sulphur-isotope composition. In all these suites the oxidation state of the granite is controlled by inheritance of the parental chemistry, as shown by their average $\delta^{34}S$ values. In some cases a smaller-scale variation is shown by extremes of oxidation state which alters the isotopic composition by loss of a fugitive component. There is no clear evidence for contamination in any of these rocks.

ACKNOWLEDGMENTS

I am indebted to Stirling Shaw who has collaborated at all stages of the project. This contribution is included with the approval of the Director, Institute of Geological Sciences.

133

REFERENCES

Adam, J.W.H. (1960). On the geology and primary tin-ore deposits in the sedimentary formation of Billiton, *Geol. Mijnbouw.*, 22, 406–426.

Agar, R.A. (1978). The Peruvian Coastal Batholith: Its monzonitic rocks and their related mineralisation, *Thesis Ph.D. University of Liverpool*, (unpubl.).

Aguirre, L. and Levi, B. (1977). Relationship between metamorphism, plutonism and geotectonics in the Palaeozoic and Mesozoic edifices of the northern segment of the Meridional Andes, In: *Plutonism in relation to Volcanism and Metamorphism* (Eds. T. Nozawa and N. Yamada), pp. 75–77; Toyama, Japan.

Aguirre, L., Levi, B. and Offler, R. (1978). Unconformities as mineralogical breaks in the burial metamorphism of the Andes, *Contrib. Mineral. Petrol.*, 66, 361–366.

Akaad, M.K. (1956). The Ardara Granitic Diapir of County Donegal, *Q.J. geol. Soc. London*, 112, 263–288.

Anderson, R.N., DeLong, S.E. and Schwarz, W.M. (1978). Termal model for subduction with dehydration in the downgoing slab, *J. Geol. Chicago*, 86, 731–739.

Aranyakanon, P. (1961). The cassiterite deposit of Haad Som Pan, Ranong Province, Thailand, *Rept. No. 4, Dept. of Mineral Resources, Bangkok.*

Amstrong, R.L., Taubeneck, W.H. and Hales, P.O. (1977). Rb-Sr and K-Ar geochronometry of Mesozoic granitic rocks and their Sr isotopic composition, Oregon, Washington and Idaho, *Bull. geol. Soc. Am.*, 88, 397–411.

Arth, J.G. and Barker, F. (1976). Rare-earth partitioning between hornblende and dacitic liquid and implications for the genesis of trondhjemitic-tonalitic magmas, *Geology*, 4, 534–536.

Arth, J.G., Barker, F., Peterman, Z.E. and Friedman, I. (1978). Geochemistry of the gabbro-diorite-tonalite-trondhjemite suite of southwest Finland and its implications for the origin of tonalitic and trondhjemitic magmas, *J. Petrol.*, 19, 289–316.

Arth, J.G. and Hanson, G.N. (1975). Geochemistry and origin of the early Precambrian crust of northeastern Minnesota, *Geochim. cosmochim. Acta*, 39, 325–362.

Atherton, M.P. and Brenchley, P.J. (1972). A preliminary study of the structure, stratigraphy and metamorphism of some contact rocks of the western Andes, near the Quebrada Venado Muerto, Peru, *Geol. J.*, 8, 161–178.

Baker, M.C.W. (1974). Volcano spacing, fractures and thickness of the lithosphere – a reply, *Earth planet. Sci. Lett.*, 23, 161–3.

Baker, M.C.W. and Francis, P.W. (1978). Upper Cenozoic volcanism in the Central Andes – ages and volumes, *Earth planet. Sci. Lett.*, 41, 175–87.

Barazangi, M. and Isacks, B.L. (1976). Spatial distribution of earthquakes and subduction of the Nazca plate beneath South America, *Geology*, 4, 686–92.

Barazangi, M. and Isacks, B.L. (1979). Subduction of the Nazca plate beneath Peru: evidence from spatial distribution of earthquakes, *Geophys. J. R. astron. Soc.*, 57, 537–55.

Barker, F. and Arth, J.G. (1976). Generation of trondhjemite – tonalite liquids and Archean bimodal trondhjemite – basalt suites, *Geology*, 4, 596–600.

Barker, F., Arth, J.G., Peterman, Z.E. and Friedman, I. (1976). The 1.7 to 1.8 b.y. – old trondhjemites of southwestern Colorado and northern New Mexico: Geochemistry and depths of genesis, *Bull. geol. Soc. Am.*, 87, 189–198.

Barker, F. and Peterman, Z.E. (1974). Bimodal tholeiitic – dacitic magmatism and the early Precambrian crust, *Precambrian Res.*, 1, 1–12.

Barker, F., Wones, D.R., Sharp, W.N. and Desbrough, G.A. (1975). The Pikes Peak batholith, Colorado Front Range, and a model for the origin of the gabbro-anorthosite-syenite-potassic granite suite, *Precambrian Res.*, 2, 97–160.

Bartlett, R.W. (1969). Magma convection, temperature distribution and differentiation, *Am. J. Sci.*, 267, 1067–1082.

Bateman, P.C. and Dodge, F.C.W. (1970). Variations of major chemical constituents across the central Sierra Nevada batholith, *Bull. geol. Soc. Am.*, 81, 409–420.

Bateman, P.C. and Chappell, B.W. (1979). Crystallization, fractionation and solidification of the Tuolumne intrusive series, Yosemite National Park, California, *Bull. geol. Soc. Am.,* 90, 465–482.

Beach, A. and Tarney, J. (1978). Major and trace element patterns established during retrogressive metamorphism of granulite facies gneisses, N.W. Scotland, *Precambrian Res.,* 7, 325–348.

Bean, J.H. (1969). Iron ore deposits of West Malaysia, *Econ. Bull. geol. Surv. West Malaysia,* 2.

Beckinsale, R.D., Suensilpong, S., Nakapadungrat, S. and Walsh, J.N. (1979). Geochronology and geochemistry of granite magmatism in Thailand in relation to a plate tectonic model, *J. geol. Soc. London,* in press.

Bell, K. (1968). Age relations and provenance of the Dalradian Series of Scotland. *Bull. geol. Soc. Am.,* 79, 1167–1194.

Best, M.G. (1975). Migration of hydrous fluids in the upper mantle and potassium variation in calc-alkalic rocks. *Geology,* 3, 429–432.

Bignell, J.D. and Snelling, N.J. (1977). Geochronology of Malayan granites. *Overseas Geol. miner. Resour. London,* No. 47.

Blatt, H. and Brown, V.M. (1974). Prophylactic separation of heavy minerals, *J. sediment. Petrol.,* 44, 260–261.

Blaxland, A.B., Aftalion, M. and van Breemen, O. (1979). Pb isotopic composition of feldspars from Scottish Caledonian granites, and the nature of the underlying crust, *Scott. J. Geol.,* 15, 139–151.

Bridgewater, D., Sutton, J. and Watterson, J. (1974). Crustal downfolding associated with igneous activity, *Tectonophysics,* 21, 57–77.

Brigue, L. and Lancelot, J.R. (1979). Rb-Sr systematics and crustal contamination models for calc-alkaline igneous rocks, *Earth planet. Sci. Lett.,* 43, 385–396.

Brooks, C. and Compston, W. (1965). The age and initial $^{87}Sr/^{86}Sr$ of the Heemskirk granite, western Tasmania, *J. geophys. Res.,* 70, 6249–6262.

Brown, G.C. (1977). Mantle origin of Cordilleran granites. *Nature. London,* 265, 21–24.

Brown, G.C. (1979a). Calc-alkaline magma genesis: The Pan-African contribution to crustal growth? In: *Proc. EMANS symposium, Jeddah;* Pergamon Press, in press.

Brown, G.C. (1979b). Geochemical and geophysical constraints on the origin and evolution of Caledonian granites. In: The Caledonides of the British Isles – reviewed (Eds. A.L. Harris, C.H. Holland and B.E. Leake), *Geol. Soc. Lond. Spec. Issue,* in press.

Brown, G.C. and Hennessy, J. (1978). The initiation and thermal diversity of granite magmatism, *Philos. Trans. R. Soc. London, A.* 288, 631–643.

Brown, G.C. and Locke, C.A. (in press). Space–time variations in British Caledonian granites: some geophysical correlations, *Earth planet. Sci. Lett.*

Brown, J.F. (1975). Rb-Sr studies and related geochemistry on the Caledonian calc-alkaline igneous rocks of N.W. Argyllshire, *Thesis D. Phil., University of Oxford* (unpubl.).

Brown, P.E., Miller, J.A. and Grasty, R.L. (1968). Isotopic ages of late Caledonian granitic intrusions in the British Isles, *Proc. Yorkshire geol. Soc.,* 36, 251–276.

Buddington, A.F. (1959). Granite emplacement with special reference to North America, *Bull. geol. Soc. Am.,* 70, 671–747.

Buma, G., Frey, F.A. and Wones, D.R. (1971). New England granites: trace element evidence regarding their origin and differentiation. *Contrib. Mineral. Petrol.,* 31, 300–320.

Burke, K., Dewey, J.F. and Kidd, W.S.F. (1976). Dominance of horizontal movements, arc and microcontinental collisions during the later permobile regime. In: *The Early History of the Earth* (Ed. B.F. Windley), pp. 113–129; Wiley Interscience, London.

Bussell, M.A. (1975). The structural evolution of the Coastal Batholith in the Provinces of Ancash and Lima, Central Peru, *Thesis Ph.D., University of Liverpool* (unpubl.).

Bussell, M.A. and McCourt, W.J. (1977). The Iglesia Irca intrusion and the role of gas brecciation in the emplacement of the Coastal Batholith of Peru, *Geol. Mag.,* 114, 375–387.

Bussell, M.A., Pitcher, W.S. and Wilson, P.A. (1976). Ring complexes of the Peruvian Batholith: a long standing sub-volcanic regime. *Can. J. Earth Sci.,* 13, 1020–30.

135

Cann, J.R. (1970). Upward movement of granite magma, *Geol. Mag.* 107, 335–40.

Carmichael, I.S.E., Turner, F.J. and Verhoogen, J. (1974). *Igneous Petrology,* McGraw Hill, New York.

Carter, S.R., Evensen, N.M., Hamilton, P.J. and O'Nions, R.K. (1978a). Continental volcanics from enriched and depleted source regions: Nd and Sr isotope evidence, *Earth planet Sci. Lett.,* 37, 401–408.

Carter, S.R., Evensen, N.M., Hamilton, P.J. and O'Nions, R.K. (1978b). Neodymium and strontium isotope evidence for crustal contamination of continental volcanics, *Science,* 202, 743–7.

Cawthorn, R.G., Strong, D.F. and Brown, P.A. (1976). Origin of corundum-normative intrusive and extrusive magmas, *Nature. London,* 259, 102–104.

Chapman, H.J. (1978). Geochronology and isotope geochemistry of Precambrian rocks from North-west Scotland, *Thesis D.Phil., University of Oxford* (unpubl.).

Chapman, H.J. and Moorbath, S. (1977). Lead isotope measurements from the oldest recognized Lewisian gneisses of north-west Scotland, *Nature. London,* 268, 41–42.

Chappell, B.W. (1978). Granitoids from the Moonbi District, New England Batholith, Eastern Australia, *J. geol. Soc. Austr.,* 25, 267–283.

Chappell, B.W. and White, A.J.R. (1974). Two contrasting granite types, *Pacific Geology,* 8, 173–174.

Chow, T.J. and Patterson, C.C. (1962). On the occurrence and significance of lead isotopes in pelagic sediments, *Geochim. cosmochim. Acta,* 26, 263–308.

Church, S.E. and Tilton, G.R. (1973). Lead and strontium isotopic studies in the Cascade mountains: bearing on andesite genesis, *Bull. geol. Soc. Am.,* 84, 431–454.

Clegg, E.L.G. (1944). Notes on tin and wolfram with a description of the tin and wolfram deposits of Burma and India, *Rec. geol. Surv. India,* Bull No. 15.

Cloos, H. (1925). *Einführung in die tektonische Behandlung magmatischer Erscheinungen (Granittektonik). I. Spezieller Teil. Das Riesengebirge in Schlesien. Bau, Bildung, und Oberflachengestältung.* 194 pp. Borntraeger, Berlin.

Cobbing, E.J. (1978). The Andean geosyncline in Peru, and its distinction from Alpine geosynclines, *J. geol. Soc. London,* 135, 207–218.

Cobbing, E.J., Baldock, J., McCourt, W.J., Pitcher, W.S., Taylor, W.P. and Wilson, J.J. The geology of the western cordillera of northern Peru, *Mem. Inst. Geol. Sci. Overseas Div.,* in press.

Cobbing, E.J., Ozard, J.M. and Snelling, N.J. (1977a). Reconnaissance geochronology of the crystalline basement rocks of the Coastal Cordillera of Southern Peru, *Bull. geol. Soc. Am.,* 88, 241–246.

Cobbing, E.J. and Pitcher, W.S. (1972). The Coastal Batholith of Central Peru, *J. geol. Soc. London,* 128, 421–460.

Cobbing, E.J., Pitcher, W.S. and Taylor, W.P. (1977b). Segments and Super-units in the Coastal Batholith of Peru, *J. Geol. Chicago,* 85, 625–631.

Coleman, M.L. (1977). Sulphur isotopes in petrology, *J. geol. Soc. London,* 133, 593–608.

Coleman, M.L. and Moore, M.P. (1978). Direct reduction of sulfates to sulfur dioxide for isotopic analysis, *Analyt. Chem.,* 50, 1594–1595.

Colony, R.J. and Sinclair, J.H. (1928). The lavas of the volcano Sumaco, Eastern Ecuador, South America, *Am. J. Sci.* 16, 299–312.

Compton, P. (1978). Rare earth evidence for the origin of the Nûk Gneisses, Buksefjorden Region, southern West Greenland, *Contrib. Mineral. Petrol.,* 66, 283–293.

Condie, K.C. (1978). Geochemistry of Proterozoic granitic plutons from New Mexico, U.S.A., *Chem. Geol.,* 21, 131–149.

Croner, A. (1976). Proterozoic crustal evolution in parts of southern Africa and evidence for extensive sialic crust since the end of the Archean. *Philos. Trans. R. Soc. London, A,* 280, 541–554.

Cumming, G.L. and Richards, J.R. (1975). Ore lead isotope ratios in a continuously changing Earth, *Earth planet. Sci. Lett.,* 28, 155–171.

136

Daly, R.A. (1933). *Igneous rocks and the depths of the earth.* 598 pp. McGraw-Hill, New York.

Davies, F.B. and Windley, B.F. (1976). The significance of major Proterozoic high-grade linear belts in continental evolution, *Nature. London,* 263, 383–385.

De Jong, K.A. (1973). Mountain building in the Mediterranean region. In: *Gravity and Tectonics* (Eds. K.A. De Jong and R. Scholten), pp. 125–139; Wiley, New York.

DePaolo, D.P. and Wasserburg, G.J. (1976). Nd isotopic variations and petrogenetic models, *Geophys. Res. Lett. Washington,* 3, 249–252.

DePaolo, D.P. and Wasserburg, G.J. (1977). The sources of island arcs as indicated by Nd and Sr isotopic studies, *Geophys. Res. Lett. Washington,* 4, 465–468.

DePaolo, D.J. and Wasserburg, G.J. (1979). Petrogenetic mixing models and Nd-Sr isotope patterns, *Geochim. cosmochim. Acta.,* 43, 615–27.

Deruelle, B. and Deruelle, J. (1974). Géologie des volcans quaternaires des Nevados de Chillán, Chili Central. *Bull. Volcanol.* 38, 425–44.

Dewey, J.F. (1969). Evolution of the Appalachian/Caledonian orogen, *Nature. London,* 222, 124–129.

Dewey, J.F. (1974). The geology of the southern termination of the Caledonides. In: *The Ocean Basins and Margins* 2, *The North Atlantic* (Eds. A.E.M. Nairn and F.G. Stehli), pp. 205–231; Plenum Press, New York.

Dewey, J.F. and Pankhurst, R.J. (1970). The evolution of the Scottish Caledonides in relation to their isotopic age pattern, *Trans. R. Soc. Edinburgh,* 68, 361–389.

de Wit, M.J. (1977). The evolution of the Scotia Arc as the key to the reconstruction of southwest Gondwanaland, *Tectonophysics,* 37, 53–81.

Dickinson, W.R. (1975). Potash-depth (K-h) relations in continental margin and intra-oceanic magmatic arcs, *Geology,* 3, 53–56.

Doe, B.R. and Delevaux, M.H. (1973). Variations in lead-isotopic composition in Mesozoic granitic rocks of California: a preliminary investigation, *Bull. geol. Soc. Am.,* 84, 3513–3526.

Dostal, J., Dupuy, C. and Léfevre, C. (1977a). Rare earth element distribution in Plio-Quaternary volcanic rocks from Southern Peru, *Lithos,* 10, 173–83.

Dostal, J., Zentilli, M., Caelles, J.C. and Clark, A.H. (1977b). Geochemistry and origin of volcanic rocks from the Andes ($26°–28°$S), *Contrib. Mineral. Petrol.,* 63, 113–28.

Drury, S.A. (1978). REE distributions in a high grade Archaean gneiss complex in Scotland: Implications for the genesis of ancient crust. *Precambrian Res.,* 7, 237–257.

Dunnet, D. (1969). A technique of finite strain analysis using elliptical particles, *Tectonophysics,* 7, 117–136.

El–Hinnawi, E.E., Pichler, H. and Zeil, W. (1969). Trace element distribution in Chilean ignimbrites, *Contrib. Mineral. Petrol.,* 24, 50–62.

Elston, W.E., Rhodes, R.C., Coney, P.J. and Deal, E.G. (1975b). Progress report on the Mogollon Plateau volcanic field, south western New Mexico, No. 3 – Surface expression of a pluton. *New Mexico Geol. Soc. Spec. Pub. No. 5,* 3–28.

Elston, W.E., Seager, W.R. and Clemons, R.E. (1975a). Emory cauldron, Black Range, New Mexico, source of the Kneeling Nun Tuff. *New Mexico Geol. Soc. Guidebook, 26th Field Conf., Las Cruces Country,* pp. 283–91.

Emslie, R.F. (1978). Anorthosite massifs, rapakivi granites, and late Proterozoic rifting of North America, *Precambrian Res.,* 7, 61–98.

Erlank, A.J. and Shimizu, N. (1977). Strontium and Sr-isotope distributions in some kimberlite nodules and minerals. (Abs), *The second International Kimberlite Conference, New Mexico.*

Eskola, P.E. (1932). On the origin of granitic magmas, *Mitt. Mineral Petrog.* 42, 455–481.

Faure, G. (1977). *Principles of Isotope Geology,* 464 pp.; Wiley, New York.

Faure, G. and Hurley, P.M. (1963). The isotopic composition of strontium in oceanic and continental basalt: Application to the origin of igneous rocks, *J. Petrol.,* 4, 31–50.

137

Faure, G. and Powell, J.L. (1972). *Strontium Isotope Geology,* 188 pp.; Springer Verlag, Berlin, Heidelberg and New York.

Fernandez, C.A., Hoermann, P.K., Kussmaul, S., Meave, J., Pichler, H. and Subieta, T. (1973). First petrologic data on young volcanic rocks of SW Bolivia, *Tschermaks Min. Petr. Mitt.,* 19, 149–72.

Flinn, D. (1965). On the symmetry principle and the deformation ellipsoid, *Geol. Mag.,* 102, 36–45.

Flood, R.H. and Shaw, S.E. (1975). A cordierite-bearing granite suite from the New England batholith, N.S.W., Australia, *Contrib. Mineral. Petrol.,* 52, 157–164.

Forester, R.W. and Taylor, H.P. (1978). $^{18}O/^{16}O$, D/H and $^{13}C/^{12}C$ studies of the Tertiary igneous complex of Skye, Scotland. *Am. J. Sci.,* 277, 163–77.

Francis, P.W. and Baker, M.C.W. (1978). Sources of two large ignimbrites in the Central Andes – some LANDSAT evidence, *J. Volcanol. Geotherm. Res.,* 4, 81–7.

Francis, P.W., Hammill, M., Kretzschmar, G. and Thorpe, R.S. (1978). The Cerro Galan Caldera, northwest Argentina and its tectonic setting, *Nature. London,* 274, 749–51.

Francis, P.W., Moorbath, S. and Thorpe, R.S. (1977). Strontium isotope data for Recent andesites in Ecuador and North Chile, *Earth planet. Sci. Lett.,* 37, 197–202.

Francis, P.W., Roobol, M.J., Walker, G.P.L., Cobbold, P.R. and Coward, M.P. (1974). The San Pedro and San Pablo volcanoes of northern Chile and their hot avalanche deposits, *Geol. Rdsch.,* 63, 357–88.

Francis, P.W. and Rundle, C. (1976). Rates of production of the main magma types in the central Andes, *Bull. geol. Soc. Am.,* 87, 474–80.

Frey, F.A., Chappell, B.W. and Roy, S.D. (1978). Fractionation of rare-earth elements in the Tuolumne intrusive series, Sierra Nevada batholith, California, *Geology,* 6, 239–242.

Frey, F.A. and Green, D.H. (1974). The mineralogy, geochemistry and origin of lherzolite inclusions in Victorian basanites. *Geochim. cosmochim. Acta.,* 38, 1023–1059.

Fryer, B.J., Fyfe, W.S. and Kerrich, R. (1979). Archaean volcanogenic oceans, *Chem. Geol.,* 24, 25–33.

Fyfe, W.S. (1978). The evolution of the Earth's crust: modern plate tectonics to ancient hot spot tectonics?, *Chem. Geol.,* 23, 89–114.

Fyfe, W.S. (1979). The geochemical cycle of uranium, *Philos. Trans. R. Soc. London, A,* 291, 433–445.

Fyfe, W.S. and McBirney, A.R. (1975). Subduction and the structure of andesitic volcanic belts, *Am. J. Sci.,* 275A, 285–97.

Gardiner, C.I. and Reynolds, S.H. (1932). The Loch Doon 'granite' area, Galloway, *Q. J. geol. Soc. London,* 88, 1–34.

Gass, I.G. (1977). The evolution of the Pan-African crystalline basement in N.E. Africa and Arabia, *J. geol. Soc. London,* 134, 129–138.

Gastil, R.G. (1975). Plutonic zones in the Peninsular Ranges of southern California and northern Baja California, *Geology,* 3, 361–363.

Ghuma, M.A. and Rogers, J.J.W. (1978). Geology, geochemistry, and tectonic setting of the Ben Ghnema batholith, Tibesti massif, southern Libya, *Bull. geol. Soc. Am.,* 89, 1351–1358.

Gibbs, R.J. (1967). The geochemistry of the Amazon River System: Part 1 – The factors that control the salinity and composition and concentration of the suspended solids, *Bull. geol. Soc. Am.,* 78, 1203–1232.

Gill, J.B. (1974). Role of underthrust oceanic crust in the genesis of a Fijian calc-alkaline suite, *Contrib. Mineral. Petrol.,* 43, 29–45.

Gill, J.B. (1978). Role of trace element partition coefficients in models of andesite genesis, *Geochim. cosmochim. Acta,,* 42, 709–724.

Green, D.H. (1975). Genesis of Archean peridotitic magmas and constraints on Archean geothermal gradients and tectonics, *Geology,* 3, 15–18.

Green, T.H., Green, D.H. and Ringwood, A.E. (1967). The origin of high-alumina basalts and their relationships to quartz-tholeiites and alkali basalts, *Earth planet. Sci. Lett.,*

138

2, 41–51.

Greenberg, J.K. (1978). Geochemistry, petrology and tectonic origin of Egyptian younger granites: *Thesis Ph.D., University of North Carolina,* (unpubl.).

Greenwood, W.R., Hadley, D.G., Anderson, R.E., Fleck, R.J. and Schmidt, D.L. (1976). Late Proterozoic cratonisation in southerwestern Saudi Arabia, *Philos. Trans. R. Soc. London, A.,* 280, 517–527.

Halliday, A.N., Aftalion, M., van Breemen, O. and Jocelyn, J. (1979). Petrogenetic significance of Rb-Sr and U-Pb isotope systems in the c. 400 Ma old British Isles granitoids and their hosts. In: *The Caledonides of the British Isles – reviewed* (Eds. A.L. Harris, C.H. Holland and B.E. Leake), Geol. Soc. Lond. Spec. Publ., in press.

Hamet, J. and Allègre, C.J. (1976). Rb-Sr systematics in granite from central Nepal (Manaslu): Significance of the Oligocene age and high $^{87}Sr/^{86}Sr$ ratio in Himalayan orogeny, *Geology,* 4, 470–472.

Hamilton, P.J., Evensen, N.M., O'Nions, R.K. and Tarney, J. (1979). Sm-Nd systematics of Lewisian gneisses: implications for the origin of granulites, *Nature. London,* 277, 25–28.

Hamilton, W. (1969). The volcanic Central Andes: A modern model for the Cretaceous batholiths and tectonics for North America, *Bull. Oregon St. Dept. Geol. Mineral. Ind.,* 65, 175–84.

Hamilton, W. and Myers, W.B. (1967). The nature of batholiths, *Prof. Pap. U.S. geol. Surv.,* 554–C, 30 pp.

Hanson, G.N. (1978). The application of trace elements to the petrogenesis of igneous rocks of granitic composition, *Earth planet. Sci. Lett.,* 38, 26–43

Harding, R.R. (1978). Some aspects of the petrography of six groups of rocks from Central Peru, *Rep. Inst. geol. Sci. London,* No. 124 (unpubl.).

Hargraves, R.B. (1976). Precambrian geologic history, *Science,* 193, 363–371.

Harris, P.G., Kennedy, W.Q. and Scarfe, C.M. (1970). Volcanism versus plutonism – The effect of chemical composition In: *Mechanism of Igneous Intrusion* (Ed. G. Newall and N. Rast), pp. 187–200, Geol. J. Spec. Issue, No. 2. Gallery Press, Liverpool.

Haslam, H.W. (1968). The crystallization of intermediate and acid magmas at Ben Nevis, Scotland. *J. Petrol.,* 9, 84–104.

Haslam, W.Q. (1946). The Great Glen Fault, *Q.J. geol. Soc. London,* 102, 41–76.

Hawkesworth, C.J., Norry, M.J., Roddick, J.C., Baker, P.E., Francis, P.W. and Thorpe, R.S. (1979a). $^{143}Nd/^{144}Nd$, $^{87}Sr/^{86}Sr$ and incompatible element variations in calc-alkaline andesites and plateau lavas from South America, *Earth planet. Sci. Lett.,* 42, 45–57.

Hawkesworth, C.J., Norry, M.J., Roddick, J.C. and Vollmer, R. (1979b). $^{143}Nd/^{144}Nd$ and $^{87}Sr/^{86}Sr$ ratios from the Azores and their significance in LIL-element enriched mantle, *Nature. London.* 280, 28–31.

Hawkesworth, C.J., O'Nions, R.K. and Arculus, R.J. (1979c). Nd- and Sr-isotope geochemistry of island arc volcanics. Grenada, Lesser Antilles, *Earth planet. Sci. Lett.,* (in press).

Hawkesworth, C.J., O'Nions, R.K., Pankhurst, R.J., Hamilton, P.J. and Evensen, N.M. (1977). A geochemical study of island-arc and back-arc tholeiites from the Scotia Sea, *Earth planet. Sci. Lett.,* 36, 253–262.

Hawkesworth, C.J. and Powell, M. (1979). The origin of andesites in the Lesser Antilles island arc, *Earth planet. Sci. Lett.,* m/s submitted.

Hawkesworth, C.J. and Vollmer, R. (1979). Crustal contamination versus enriched mantle: $^{143}Nd/^{144}Nd$ and $^{87}Sr/^{86}Sr$ evidence from the Italian volcanics, *Contrib. Mineral. Petrol.* 69, 151–169.

Herbert, H.K. and Smith, J.W. (1978). Sulfur isotopes and origin of some sulfide deposits, New England, Australia, *Mineral. Deposita Berlin,* 13, 51–63.

Hine, R., Williams, I.S., Chappell, B.W. and White, A.J.R. (1978). Contrasts between I- and S-type granitoids of the Kosciusko batholith, *J. geol. Soc. Austr.,* 25, 219–234.

Hobson, G.V. (1940). The development of the mineral deposit at Mawchi as determined by geology and genesis, *Trans. Ming. geol. metall. Inst. India,* 36, 35–78.

Hodge, D.S. (1974). Thermal models for the origin of granitic batholiths, *Nature. London,* 251, 297–299.

139

Hoffman, P. (1973). Evolution of an early Proterozoic continental margin: the Coronation Geosyncline and associated aulacogens of the north-western Canadian Shield, *Philos. Trans. R. Soc. London, A,* 273, 547–581.

Høgdahl, O.T., Bowen, B.T. and Melson, S. (1968). Neutron activation analysis of lanthanide elements in sea water, *Advan. Chem. Ser.,* 73, 308–325.

Holland, C.H., Kelling, G. and Walton, E.K. (in press). O.T. Jones and after; a multitude of models. In: *The Caledonides of the British Isles – reviewed,* (Eds. A.L. Harris, C.H. Holland and B.E. Leake), Geol. Soc. Lond. Spec. Publ.

Hosking, K.F.G. (1969). Aspects of the geology of the tin-fields of Southeast Asia, In: *Second Technical Conference on Tin, Bangkok, Thailand,* 1, 39–80. The International Tin Council.

Hosking, K.F.G. (1973). Primary Mineral Deposits. In: *Geology of the Malay Peninsula, West Malaysia and Singapore.* (Eds. D.J. Gobbet and C.S. Hutchison), pp. 335–90; Wiley-Interscience, New York.

Hunter, D.R., Barker, F. and Millard, H.T. (1978). The geochemical nature of the Archean Ancient Gneiss Complex and Granodiorite Suite, Swaziland: a preliminary study, *Precambrian Res.,* 7, 105–107.

Hutchinson, W.W. (1970). Metamorphic framework and plutonic styles in the Prince Rupert region of the central Coast Mountains, British Columbia, *Can. J. Earth Sci.,* 7, 376–405.

Hutchison, C.S. and Taylor, D. (1978). Metallogenesis in S.E. Asia, *J. geol. Soc. London,* 135, 407–428.

Ishihara, S. (1977). The magnetite-series and ilmenite-series granitic rocks. *Ming. Geol.,* 27, 293–305.

Ishihara, S. (1978). Metallogenesis in the Japanese island-arc system, *J. geol. Soc. London,* 135, 389–406.

Ivanova, G.F. and Butuzova, Ye.G. (1968). Distribution of tungsten, tin and molybdenum in the granites of Eastern Transbaykalia, *Geochem. Int.,* 5, 572–583.

Jakeš, P. and White, A.J.R. (1972). Major and trace element abundances in volcanic rocks and orogenic areas, *Bull. geol. Soc. Am.,* 83, 29–40.

James, D.E. (1971). Plate Tectonic Model for the evolution of the Central Andes, *Bull. geol. Soc. Am.,* 82, 3325–46.

James, D.E., Brooks, C. and Cuyubamba, A. (1976). Andean Cenozoic volcanism: magma genesis in the light of strontium isotopic composition and trace element geochemistry, *Bull. geol. Soc. Am.,* 87, 592–600.

Kay, R.W., Sun, S.S. and Lee-hu, C.N. (1978). Pb and Sr isotopes in volcanic rocks from the Aleutian Islands and Pribilof Islands, Alaska, *Geochim. cosmochim. Acta.,* 42, 263–273.

Kennedy, W.Q. (1946). The Great Glen Fault, *Q. J. geol. Soc. London,* 102, 41–76.

Kennett, J.P., McBirney, A.R. and Thunell, R.C. (1977). Episodes of Cenozoic volcanism in the Circum-Pacific region, *J. Volcanol. Geotherm. Res.,* 2, 145–163.

Khin Zaw, U. (1978). Fluid inclusion studies on the Hermingyi tungsten-tin deposit, southern Burma. *Proc. Third Regional Conference on Geology and mineral resources of Southeast Asia,* 393–400.

King, R.F. (1966). The magnetic fabric of some Irish granites, *Geol. J.,* 5, 43–66.

Kistler, R.W. (1974). Phanerozoic batholiths in Western North America: a summary of some recent work on variations in time, space, chemistry and isotopic composition, *Ann. Rev. Earth planet. Sci.,* 2, 404–418.

Kistler, R.W. and Peterman, Z.E. (1973). Variations in Sr, Rb, K, Na and initial Sr^{87}/Sr^{86} in Mesozoic granitic rocks and intruded wall rocks in central California. *Bull. geol. Soc. Am.,* 84, 3489–3512.

Klerkx, J., Deutsch, S., Pichler, H. and Zeil, W. (1977). Strontium isotope composition and trace element data bearing on the origin of Cenozoic volcanic rocks of the central and southern Andes, *J. Volcanol. Geotherm. Res.,* 2, 49–71.

Knox, G.J. (1974). The structure and emplacement of the Rio Fortaleza centred acid complex, Ancash, Peru, *J. geol. Soc. London,* 130, 295–308.

Kramers, J.D. (1977). Lead and strontium isotopes in Cretaceous kimberlites and mantle-derived xenoliths from Southern Africa, *Earth planet. Sci. Lett.*, 34, 419–431.

Kröner, A. (1976). Proterozoic crustal evolution in parts of southern Africa and evidence for extensive sialic crust since the end of the Archaean, *Philos. Trans. R. Soc. London, A.,* 280, 541–554.

Kushiro, I. (1972). Effect of water on the composition of magmas formed at high pressures, *J. Petrol.,* 13, 311–334.

Kussmaul, S., Kormann, P.K., Ploskonka, E. and Subieta, T. (1977). Volcanism and structure of south western Bolivia, *J. Volcanol. Geotherm. Res.,* 2, 73–111.

Lambert, I.B. and Wyllie, P.J. (1972). Melting of gabbro (quartz eclogite) with excess water to 35 kilobars, with geological implications, *J. Geol. Chicago,* 80, 693–708.

Larsen, E.S. (1948). Batholith and associated rocks of Corona, Elsinore and San Luis Rey quadrangles, Southern California, *Mem. geol. Soc. A.,* 29, 182 pp.

Leake, B.E. (1978). Granite emplacement: the granites of Ireland and their origin. In: Crustal evolution in northwest Britain and adjacent regions (Eds. D.R. Bowes and B.E. Leake), pp. 221–248; *Geol. J. Special Issue,* No. 10.

Lefévre, C. (1973). Les caractères magmatiques du volcanisme plioquaternaire des Andes dans le Sud du Pérou, *Contrib. Mineral. Petrol.,* 41, 259–72.

Le Maitre, R.W. (1976). The chemical variability of some common igneous rocks, *J. Petrol.,* 17, 589–637.

Levinson, A.A. (1974). *Introduction to exploration geochemistry.* 612 pp. Applied Publishing, Calgary.

Lloyd, F. E. and Bailey, D.K. (1975). Light element metasomatism of the continental mantle: the evidence and consequences, *Phys. Chem. Earth.,* 9, 389–416.

Lopez-Escobar, L. and Frey, F.A. (1976). Rocas volcanicas cuaternarieas de Chile central-sur $(33^{\circ}-41^{\circ}S)$: Modelas petrogeneticos sugeridos por las Tierras Raras, *Actas I Congresso Geologico Chileno, Dep. Geol. Univ. Chile,* Tomo II, F223–F225.

Lopez-Escobar, L., Frey, F.A. and Vergara, M. (1976). Andesites from central-south Chile: trace element abundances and petrogenesis. In: *Proceedings of the Symposium on Andean and Antarctic Volcanology Problems,* (Santiago, Chile, 1974), (Ed. O. Gonzales-Ferran). pp. 725–61. Giannini and Figli, Naples.

Lopez-Escobar, L., Frey, F.A. and Vergara, M. (1977). Andesites and high-alumina basalts from central-south Chile High Andes: geochemical evidence bearing on their petrogenesis. *Contrib. Mineral. Petrol.,* 63, 199–228.

MacDonald, G.A. (1972). *Volcanoes,* Prentice-Hall, Inc., New Jersey.

MacLeod, W.N., Turner, D.C. and Wright, E.P. (1971). The geology of the Jos Plateau, Vol. 1, general geology, *Bull. geol. Surv. Nigeria,* 32, 110 pp.

Marsh, B.D. (1976). Some Aleutian andesites: their nature and source, *J. Geol. Chicago,* 84, 27–45.

Marston, R.J. (1971). The Foyers granitic complex, Inverness-shire, Scotland, *Q. J. geol. Soc. London,* 126, 331–368.

Masuda, A., Nakamura, N. and Tanaka, T. (1973). Fine structures of mutually normalised rare-earth patterns of chondrites. *Geochim. cosmochim. Acta,* 37, 239–248.

McCourt, W.J. (1978). The geochemistry and petrogenesis of the Coastal Batholith of Peru, Lima segment, *Thesis Ph.D., University of Liverpool,* (unpubl.).

McCulloch, M.T. and Wasserburg, G.J. (1978). Sm-Nd and Rb-Sr chronology of continental crust formation, *Science,* 200, 1003–11.

McNutt, R.H., Crockett, J.H., Clark, A.H., Caelles, J.C., Farrar, E., Haynes, S.J. and Zentilli, M. (1975). Initial $^{87}Sr/^{86}Sr$ ratios of the plutonic and volcanic rocks of the central Andes between latitudes 26° and 29° South, *Earth planet. Sci. Lett.,* 27, 305–323.

Meade, R.H., Nordin, C.F., Curtis, W.F., Rodrigues, F.M.C., DoVale, C.M. and Edmond, J.M. (1979). Sediment loads in the Amazon River, *Nature. London,* 278, 161–163.

Meijer, A. (1976). Pb and Sr isotopic data bearing on the origin of volcanic rocks from the Mariana island arc system, *Bull. geol. Soc. Am.,* 87, 1358–1369.

141

Mercy, E.L.P. (1963). The geochemistry of some Caledonian granitic and metasedimentary rocks. In: *The British Caledonides* (Eds. M.R.W. Johnson and F.H. Stewart), pp. 189–215; Oliver and Boyd, Edinburgh and London.

Mitchell, A.H.G. (1977). Tectonic settings for emplacement of southeast Asian tin granites, *Bull. geol. Soc. Malaysia, 9,* 123–140.

Miyashiro, A. (1961). Evolution of metamorphic belts, *J. Petrol., 2,* 277–311.

Miyashiro, A. (1967). Orogeny, regional metamorphism, and magmatism in the Japanese Islands, *Medd. Dansk geol. Foren, 17,* 390–446.

Miyashiro, A. (1973). Paired and unpaired metamorphic belts, *Tectonophysics, 17,* 241–254.

Moorbath, S. (1975). Evolution of Precambrian crust from strontium isotopic evidence, *Nature. London, 254,* 395–398.

Moorbath, S. (1977). Ages, isotopes and evolution of Precambrian continental crust, *Chem. Geol., 20,* 151–187.

Moorbath, S. (1978). Age and isotopic evidence for the evolution of continental crust, *Philos. Trans. R. Soc. London, A, 288,* 401–412.

Moorbath, S. and Bell, J.D. (1965). Strontium isotope abundance studies and rubidium-strontium determinations on Tertiary igneous rocks from the Isle of Skye, northwest Scotland, *J. Petrol., 6,* 37–66.

Moorbath, S. and Thompson, R.N. (in press). Strontium isotope geochemistry and petrogenesis of the early Tertiary lava pile of the Isle of Skye, Scotland and other basic rocks of the British Tertiary Province, *J. Petrol.*

Moorbath, S., Thorpe, R.S. and Gibson, I.L. (1978). Strontium isotope evidence for petrogenesis of Mexican andesites, *Nature. London, 271,* 437–439.

Moorbath, S. and Welke, H. (1968). Lead isotope studies on igneous rocks from the Isle of Skye, northwest Scotland, *Earth planet. Sci. Lett., 5,* 217–30.

Moreno, R.H. (1976). The Upper Cenozoic volcanism in the Andes of southern Chile (from $40^\circ\, 00'$ to $41^\circ\, 30'$ S.L.). In: *Proceedings of the Symposium on Andean and Antarctic Volcanology Problems* (Santiago, Chile, 1974) (Ed. O. Gonzales-Ferran). pp. 133–71. Giannini and Figli, Naples.

Mullen, H.S. and Bussell, M.A. (1977). The basic rock series in batholithic association, *Geol. Mag., 114,* 265–280.

Myers, J.S. (1975a). Cauldron subsidence and fluidisation: mechanisms of intrusion of the Coastal Batholith of Peru into its own volcanic ejecta, *Bull. geol. Soc. Am., 86,* 1209–1220.

Myers, J.S. (1975b). Vertical crustal movements of the Andes in Peru, *Nature. London, 254,* 672–4.

Mysen, B.O. (1978). The role of descending plates in the formation of andesite melts beneath island arcs. *Carnegie Inst. of Washington Yearbook 77,* (1977–1978), 797–801.

Neary, C.R., Gass, I.G. and Cavanagh, B.J. (1976). Granitic association of northeastern Sudan, *Bull. geol. Soc. Am., 87,* 1501–1512.

Noble, D.C., Bowman, H.R., Herbert, A.J., Silberman, M.L., Heropoulous, C.E., Fabbi, B.P., and Hedge, C.E. (1975). Chemical and isotopic constraints on the origin of low-silica latite and andesite from the Andes of central Peru, *Geology, 3,* 501–520.

Noble, D.C., McKee, E.H., Farrar, E. and Peterman, U. (1974). Episodic Cenozoic volcanism and tectonism in the Andes of Peru, *Earth planet. Sci. Lett., 21,* 213–220.

Nockolds, S.R. (1941). The Garabal Hill-Glen Fyne igneous complex, *Quart. J. geol. Soc. Lond., 96,* 451–508.

Nockolds, S.R. (1940). The Garabal Hill-Glen Fyne igneous complex, *Quart. J. geol. Soc.* rocks: a study in the relationships between the major and trace elements of igneous rocks and their minerals, *Trans. R. Soc. Edinburgh, 61,* 533–575.

Norry, M.J., Truckle, P.H., Lippard, S.J., Hawkesworth, C.J., Weaver, S.D. and Marriner, G.F. (1979). Isotopic and trace element evidence from lavas, bearing on mantle heterogeneity beneath Kenya, *Philos. Trans. R. Soc. London,* (in press).

Oba, N. (1977). Emplacement of granitic rocks in the outer zone of southwest Japan and geological significance, *J. Geol. Chicago, 85,* 383–393.

142

O'Connor, P.J. and Brück, P.M. (1976). Strontium isotope ratios for some Caledonian igneous rocks from central Leinster, Ireland, *Geol. Surv. Ireland Bull.*, 2, 69–77.

O'Neil, J.R. and Chappell, B.W. (1977). Oxygen and hydrogen isotope relations in the Berridale batholith, *J. geol. Soc. London*, 133, 559–571.

O'Neil, J.R., Shaw, S.E. and Flood, R.H. (1977). Oxygen and hydrogen isotope compositions as indicators of granite genesis in the New England batholith, Australia, *Contrib. Mineral. Petrol.*, 62, 313–328.

O'Nions, R.K., Carter, S.R., Cohen, R.S., Evensen, N.M. and Hamilton, P.J. (1978). Pb, Nd and Sr iotopes in oceanic ferromanganese depostis and ocean floor basalts, *Nature. London*, 273, 435–438.

O'Nions, R.K., Hamilton, P.J. and Evensen, N.M. (1977). Variations in $^{143}Nd/^{144}Nd$ and $^{87}Sr/^{86}Sr$ ratios in oceanic basalts, *Earth planet. Sci. Lett.*, 34, 13–22.

O'Nions, R.K. and Pankhurst, R.J. (1978). Early Archaean rocks and geochemical evolution of the Earth's crust, *Earth planet. Sci. Lett.*, 38, 211–236.

Pankhurst, R.J. (1974). Rb-Sr whole-rock chronology of Caledonian events in northeast Scotland, *Bull. geol. Soc. Am.*, 85, 345–350.

Pankhurst, R.J. (1977a). Open system crystal fractionation and incompatible element variation in basalts, *Nature. London*, 268, 36–38.

Pankhurst, R.J. (1977b). Strontium isotope evidence for mantle events in the continental lithosphere, *J. geol. Soc. London*, 134, 255–268.

Pankhurst, R.J. and Pidgeon, R.T. (1976). Inherited isotope systems and the source region pre-history of early Caledonian granites in the Dalradian Series of Scotland, *Earth planet. Sci. Lett.*, 31, 55–68.

Parslow, G.R. (1968). The physical and structural features of the Cairnsmore of Fleet granite and its aureole, *Scott. J. Geol.*, 4, 91–108.

Peacock, M.A. (1931). Classification of Igneous Rock Series, *J. Geol. Chicago*, 39, 54–67.

Pearce, J.A. (1976). Porphyry Copper Case Study, In: *Science – A Third-Level Course, Earth Science Topics and Methods* (Ed. F. Aprahamiam). The Open University Press, Milton Keynes.

Pearce, J.A. and Cann, J.R. (1973). Tectonic setting of basic volcanic rocks determined by using trace element analyses, *Earth planet. Sci. Lett.*, 19, 290–300.

Phillips, W.E.A., Stillman, C.J. and Murphy, T. (1976). A Caledonian plate tectonic model, *J. geol. Soc. London*, 132, 579–609.

Phillips, W.J. (1956). The Criffell-Dalbeattie granodiorite complex, *Q.J. geol. Soc. London*, 112, 221–239.

Pichler, H., Horman, P.K. and Braun, A.F. (1976). First petrologic data on lavas of the volcano El Reventador (Eastern Ecuador), *Munster. Forsch. Geol. Palaeont.*, 38/39, 129–41.

Pichler, H. and Zeil, W. (1972). The Cenozoic rhyolite-andesite associations of the Chilean Andes, *Bull. Volcanol.* 35, 424–52.

Pidgeon, R.T. and Aftalion, M. (1978). Cogenetic and inherited zircon U-Pb systems in granites: Palaeozoic granites of Scotland and England. In: *Crustal evolution in northwestern Britain and adjacent regions* (Eds. D.R. Bowes and B.E. Leake), pp. 183–248; Geol. J. Spec. Issue No. 10.

Pidgeon, R.T. and Johnson, M.R.W. (1974). A comparison of zircon U-Pb and whole-rock Rb-Sr systems in three phases of the Carn Chuinneag Granite, northern Scotland, *Earth planet. Sci. Lett.*, 24, 105–112.

Piper, J.D.A. (1976). Palaeomagnetic evidence for a Proterozoic super-continent, *Philos. Trans. R. Soc. London, A,* 280, 469–489.

Pitakpaivan, K. (1969). Tin bearing granite and tin barren granite in Thailand. In: *Second Technical Conference on Tin, Bangkok, Thailand,* 1, 283–298.

Pitcher, W.S. (1975). On the rate of emplacement of batholiths, *J. geol. Soc. London,* 131, 587–591.

Pitcher, W.S. (1978). The anatomy of a batholith, *J. geol. Soc. London,* 135, 157–182.

Pitcher, W.S. (1979). A commentary on the nature, ascent and emplacement of granitic magmas, *J. geol. Soc. London* (in press).

Pitcher, W.S. and Berger, A.R. (1972). *The geology of Donegal: a study of granite emplacement and unroofing.* 435 pp., Wiley-Interscience, London.

Pitcher, W.S. and Bussell, M.A. (1977). Structural control of batholith emplacement in Peru: a review, *J. geol. Soc. London,* 133, 249–256.

Ponce, R.F. (1976). El Volcan Hudson. In: *Proceedings of the Symposium on Andean and Antarctic Volcanology Problems* (Santiago, Chile, 1974), (Ed. O. Gonzalez-Ferran), pp. 78–87, Giannini and Figli, Naples.

Pongsapich, W. and Mahawat, C. (1977). Some aspects of Tak Granites, northern Thailand, *Bull. geol. Soc. Malaysia,* 9, 175–186.

Presnall, D.C. and Bateman, P.C. (1973). Fusion relations in the system $NaAlSi_3O_8 - CaAl_2Si_2O_8 - KAlSi_3O_8 - SiO_2 - H_2O$ and generation of granite magmas in the Sierra Nevada batholith, *Bull. geol. Soc. Am.,* 84, 3181–3202.

Priem, H.N.A., Boelrijk, N.A.I.M., Bon, E.H., Hebeda, E.H., Verdurman, E.A.Th. and Vershure, R.H. (1975). Isotope geochronology in the Indonesian Tin Belt, *Geol. Mijnbouw,* 54, 61–70.

Read, H.H. (1951). Metamorphism and granitization. Alex. L. du Toit Memorial Lecture, No. 2, *Trans. geol. Soc. S. Africa,* Annex. 54, 1–17.

Read, H.H. (1957). *The granite controversy,* T. Murby, London.

Read, H.H. (1961). Aspects of Caledonian magmatism in Britain, *Liverpool Manchester geol. J.,* 2, 653–683.

Regan, P.F. (1976). The genesis and emplacement of mafic plutonic rocks of the Coastal Andean Batholith, Lima Province, Peru, *Thesis Ph.D., University of Liverpool,* (unpubl.):

Ries, A.C. (1977). Rb/Sr ages from the Arequipa massif, southern Peru, *20th Ann. Rep. Res. Inst. Afr. Geol., Univ. of Leeds,* 74–77.

Ringwood, A.E. (1974). The petrological evolution of island arc systems, *J. geol. Soc. London,* 130, 183–204.

Ringwood, A.E. (1977). Petrogenesis in island arc systems. In *Island Arcs, Deep Sea Trenches and Back-arc Basins* (Eds. M. Talwani and W.C. Pitman) Maurice Ewing Series, 1, pp. 311–324; American Geophysical Union, Washington.

Robinson, B.W. and Kusakabe, M. (1975). Quantitive preparation of sulfur dioxide for $^{34}S/^{32}S$ analyses, from sulfides by combustion with cuprous oxide, *Analyt. Chem.,* 47, 1179–1181.

Roobol, M.J., Francis, P.W., Ridley, W.I., Rhodes, M. and Walker, G.P.L. (1976). Physicochemical characters of the Andean volcanic chain between latitudes $21°$ and $22°$ south. In: *Proceedings of the Symposium on Andean and Antarctic Volcanology Problems* (Santiago, Chile, 1974), (Ed. O. Gonzalez-Ferran), pp. 450–464, Giannini and Figli, Naples.

Rutherford, N.F. and Heming, R.F. (1978). The volatile component of Quaternary ignimbrite magmas from the North Island, New Zealand. *Contrib. Mineral. Petrol.,* 65, 401–11.

Sabine, P.A. (1963). The Strontian granite complex, Argyllshire, *Bull. geol. Surv. G.B.,* 20, 6–42.

Sakuyama, M. (1979). Lateral variations of H_2O contents in Quaternary magmas of Northeastern Japan, *Earth planet. Sci. Lett.,* 43, 103–11.

Sasaki, A. and Ishihara, S. (1979). Sulfur isotopic composition of the magnetite-series and ilmenite-series granitoids in Japan, *Contrib. Mineral. Petrol.,* 68, 107–115.

Saunders, A.D., Tarney, J., Stern, C.R. and Dalziel, I.W.D. (1979). Geochemistry of Mesozoic marginal basin floor igneous rocks from southern Chile, *Bull. geol. Soc. Am.,* 90, 237–258.

Saunders, A.D., Tarney, J. and Weaver, S.D. (in press b). Transverse geochemical variations across the Antarctic Peninsula: implications for calc-alkaline magma genesis, *Earth planet. Sci. Lett.*

Saunders, A.D., Weaver, S.D. and Tarney, J. (in press a). The pattern of Antarctic Peninsula plutonism. In: *Antarctic Geoscience* (Ed. C. Craddock), Univ. Wisconsin Press, Madison.

144

Schneider, A. (1970). The sulfur isotope composition of basaltic rocks, *Contrib. Mineral. Petrol.*, 25, 95–124.

Sclater, J.G. and Francheteau, J. (1970). The implications of terrestrial heat flow observations on current tectonic and geochemical models of the crust and upper mantle of the Earth. *Geophys. J.R. astron. Soc.*, 20, 509–542.

Shackleton, R.M., Ries, A.C., Coward, M.P. and Cobbold, P.R. (1979). Structure, metamorphism and geochronology of the Arequipa Massif of coastal Peru, *J. geol. Soc. London*, 136, 195–214.

Shaw, D.M., Dostal, J. and Keays, R.R. (1976). Additional estimates of continental Surface Precambrian Shield composition in Canada, *Geochim. cosmochim. Acta*, 40, 73–83.

Shaw, H.R. (1965). Comments on viscosity and crystal settling and convection in granite magmas, *Am. J. Sci.*, 263, 120–152.

Sheraton, J.W., Skinner, A.C. and Tarney, J. (1973). The geochemistry of the Scourian gneisses of the Assynt district. In: *The Early Precambrian of Scotland and Related Rocks of Greenland* (Eds. R.G. Park and J. Tarney). pp. 13–30; University of Keele.

Shilo, N.A. and Milov, A.P. (1977). Late Mesozoic granitic magmatism in the geological structures of the U.S.S.R. North-East, *Bull. geol. Soc. Malaysia*, 9, 117–122.

Sillitoe, R.H. (1974). Tectonic segmentation of the Andes: implications for magmatism and metallogeny, *Nature. London*, 250, 542–545.

Silver, L.T. and Deutsch, S. (1963). Uranium-lead isotopic variations in zircons: A case study, *J. Geol. Chicago*, 71, 721–758.

Simpson, P.R., Brown, G.C., Plant, J. and Ostle, D. (1979). Uranium mineralization and magmatism in the British Isles, *Philos. Trans. R. Soc. London, A*, 291, 133–160.

Smith, A.G. and Haliam, A. (1970). The fit of the southern continents, *Nature. London*, 225, 139–144.

Snelling, N.J., Hart, R. and Harding, R.R. (1970). Age determination on samples from the Phuket region of Thailand, *Rep. Inst. geol. Sci. London*, No. IGU 70.19 (unpubl.).

Stephansson, O. (1975). Polydiapirism of granitic rocks in the Svecofenrian of Central Sweden, *Precambrian Res.*, 2, 189–214.

Stephens, W.E. and Halliday, A.N. (in prep.). Discontinuities in the composition surface of a zoned pluton, Criffell, Scotland.

Stern, C.R. (1974). Melting products of olivine tholeiite basalt in subduction zones, *Geology*, 2, 227–30.

Stern, C.R. and Stroup, J.B. (in press). Geochemistry of the Patagonian Batholith between 51°S and 52°S latitude. In: *Antarctic Geoscience* (Ed. C. Craddock). Univ. Wisconsin Press, Madison.

Stern, C.R. and Wyllie, P.J. (1978). Phase compositions through crystallization intervals in basalt-andesite-H_2O at 30 kbar with implications for subduction zone magmas, *Am. Mineral.*, 63, 641–63.

Suensilpong, S., Meesook, A., Nakapadungrat, S. and Putthapiban, P. (1977). The granitic rocks and mineralization of the Khuntan Batholith, Lampang, *Bull. geol. Soc. Malaysia*, 9, 159–173.

Sun, S.S. and Nesbitt, R.W. (1977). Chemical heterogeneity of the Archaean mantle, composition of the bulk earth and mantle evolution, *Earth planet. Sci. Lett.*, 35, 429–448.

Tarney, J. (1973). The Scourie dyke suite and the nature of the Inverian event in Assynt. In *The Early Precambrian of Scotland and related rocks of Greenland* (Eds. R.G. Park and J. Tarney), pp. 105–118; University of Keele.

Tarney, J. (1976). Geochemistry of Archaean high grade gneisses, with implications as to the origin and evolution of the Precambrian crust. In: *The Early History of the Earth* (Ed. B.F. Windley). pp. 405–417; Wiley, London.

Tarney, J., Dalziel, I.W.D. and de Wit, M.J. (1976). Marginal basin 'Rocas Verdes' complex from S. Chile: a model for Archaean greenstone belt formation. In: *The Early History of the Earth* (Ed. B.F. Windley). pp. 131–146; Wiley Interscience, London.

Tarney, J., Saunders, A.D. and Weaver, S.D. (1977). Geochemistry of volcanic rocks from the island arcs and marginal basins of the Scotia Arc region. In: *Island Arcs, Deep Sea Trenches and Back-arc Basins* Eds. M. Talwani and W.C. Pitman), Maurice Ewing Series, 1, pp. 367–377; American Geophysical Union, Washington.

Tarney, J., Saunders, A.D., Weaver, S.D., Donnellan, N.C.B. and Hendry, G.L. (1978). Minor element geochemistry of basalts from Leg 49, North Atlantic Ocean, *Init. Repts. Deep Sea Drilling Project*, 49, 657–691.

Tarney, J., Skinner, A.C. and Sheraton, J.W. (1972). A geochemical comparison of major Archaean gneiss units from Northwest Scotland and East Greenland, *Rept. 24th Int. geol. Congress, Montreal*, 1, 162–174.

Tarney, J., Weaver, B. and Drury, S.A. (1979). Geochemistry of Archaean trondhjemitic and tonalitic gneisses from Scotland and East Greenland. In: *Trondhjemites, dacites and related rocks* (Ed. F. Barker). pp. 275–299; Elsevier, Amsterdam.

Tarney, J. and Windley, B.F. (1977). Chemistry, thermal gradients and evolution of the lower continental crust, *J. geol. Soc. London*, 134, 153–172.

Tarney, J., Wright, A.E., Gibson, I.L. and Marriner, G.F. (in prep.) Geochemistry of a Proterozoic high-K plutonic complex from East Greenland.

Tarney, J., Wood, D.A., Saunders, A.D., Varet, J. and Cann, J.R. (in press). Nature of mantle heterogeneity in the North Atlantic: evidence from Leg 49. In: *Implications of Deep Drilling Results in the Atlantic Ocean*, Maurice Ewing Ser. 2, Am. Geophys. Union, Washington.

Tauson, L.V. and Kozlov, V.D. (1973). Distribution functions and ratios of trace-element concentrations as estimators of the ore-bearing potential of granites. Cited in Levinson, A.A. (1974), *Introduction to exploration geochemistry*, p. 318, Applied Publishing, Calgary.

Taylor, H.P. (1974). The application of oxygen and hydrogen isotope studies to problems of hydrothermal alteration and ore deposition, *Econ. Geol.*, 69, 843–883.

Taylor, H.P. (1977). Water/rock interactions and the origin of water in granitic batholiths, *J. geol. Soc. London*, 133, 509–558.

Taylor, W.P. (1976). Intrusion and differentiation of granitic magma at a high level in the crust: the Puscao pluton, Lima Province, Peru, *J. Petrol.*, 17, 194–218.

Teggin, D.E. (1975). The granites of northern Thailand, *Thesis Ph.D., University of Manchester* (unpubl.).

Thode, H.G., Monster, J. and Dunford, H.B. (1961). Sulphur isotope geochemistry, *Geochim. cosmochim. Acta*, 45, 159–174.

Thorpe, R.S. and Francis, P.W. (1979). Variations in Andean andesite compositions and their petrogenetic significance, *Tectonophysics*, 57, 53–70.

Thorpe, R.S., Francis, P.W. and Moorbath, S. (1979). Rate earth and strontium isotope evidence concerning the petrogenesis of North Chilean ignimbrites, *Earth planet, Sci. Lett.*, 42, 359–67.

Thorpe, R.S., Potts, P.J. and Francis, P.W. (1976). Rare-earth data and petrogenesis of andesites from the North Chilean Andes, *Contrib. Mineral. Petrol.*, 54, 65–78.

Thorpe, R.S., Potts, P.J. and Sarre, M.B. (1977). Rare-earth evidence concerning the origin of granites of the Isle of Skye, Northwest Scotland, *Earth planet. Sci. Lett.*, 36, 111–120.

Turi, B. and Taylor, H.P. (1976). Oxygen isotope studies of potassic volcanic rocks of the Roman province, central Italy, *Contrib. Mineral. Petrol.*, 55, 1–31.

Van Breemen, O., Aftalion, M. and Johnson, M.R.W. (in press). Age of the Loch Borolan complex and late movements along the Moine Thrust, *Scott. J. Geol.*

Van Breemen, O., Aftalion, M., Pankhurst, R.J. and Richardson, S.W. (1979). Age of the Glen Dessary syenite. Inverness-shire: diachronous Palaeozoic metamorphism across the Great Glen, *Scott. J. Geol.*, 15, 49–62.

Van Breemen, O., Hutchinson, J. and Bowden, P. (1975). Age and origin of the Nigerian Mesozoic granites: a Rb-Sr study, *Contrib. Mineral. Petrol.*, 50, 157–172.

Vance, J.A. (1961). Zoned granitic intrusions – An alternative hypothesis of origin, *Bull. geol. Soc. Am.*, 72, 1723–1728.

146

Vergara, M. (1972). Note on the zonation of the Upper Cenozoic volcanism of the Andean area of central-south Chile and Argentina. In: *Symposium on the results of upper mantle investigations with emphasis on Latin America.* pp. 381–97; International Upper Mantle Project, Buenos Aires.

Von Braun, E., Besang, C., Eberle, W., Harre, W., Kreuzer, H., Lenz, H., Muller, P. and Wendt, I. (1976). Radiometric Age Determinations of Granites in Northern Thailand, *Geol. Jahrbuch. Hannover*, B21, 171–204.

Wadge, A.J., Gale, N.H., Beckinsale, R.D. and Rundle, C.C. (1978). A Rb-Sr isochron for the Shap granite, *Proc. Yorkshire geol. Soc.*, 42, 297–305.

Wager, L.R., Vincent, E.A., Brown, G.M. and Bell, J.D. (1965). Marscoite and related rocks of the Western Red Hills complex, Isle of Skye, *Philos. Trans. R. Soc. London, A*, 257, 273–308.

Walsh, J.N., Beckinsale, R.D., Skelhorn, R.R. and Thorpe, R.S. (in press). Geochemistry and petrogenesis of Tertiary granite rocks from the Island of Mull, Northwest Scotland.

Walton, M. (1955). The emplacement of granite, *Am. J. Sci.*, 253, 1–18.

Watson, J.V. (1964). Conditions in the metamorphic Caledonides during the period of late-orogenic cooling, *Geol. Mag.*, 101, 457–465.

Watson, J.V. and Plant, J. (1979). Regional geochemistry of uranium as a guide to deposit formation. *Philos. Trans. R. Soc. London, A*, 291, 321–338.

Weaver, B.L. and Tarney, J. (1979). Thermal aspects of komatiite generation and greenstone belt models, *Nature. London*, 279, 689–692.

Weaver, B.L., Tarney, J., Windley, B.F., Sugavanam, E.B. and Venkata-Rao, V. (1978). Madras granulites: geochemistry and P-T conditions of crystallisation. In: *Archaean Geochemistry* (Eds. B.F. Windley and S.M. Naqvi). pp. 177–204; Elsevier, Amsterdam.

Weaver, S.D. (1976). The Quaternary caldera volcano Emuruangogolak, Kenya Rift, and the petrology of a bimodal ferrobasalt-pantelleritic trachyte association, *Bull. Volcanol.*, 40, 209–230.

Weaver, S.D., Saunders, A.D., Pankhurst, R.J. and Tarney, J. (1979). A geochemical study of magmatism associated with the initial stages of back-arc spreading: the Quaternary volcanics of Bransfield Strait, from South Shetland Islands, *Contrib. Mineral. Petrol.*, 68, 151–169.

Weaver, S.D., Saunders, A.D. and Tarney, J. (in press). Mesozoic-Cenozoic volcanism in the South Shetland Islands and the Antarctic Peninsula: geochemical nature and plate tectonic significance. In: *Antarctic Geoscience* (Ed. C. Craddock), Univ. Wisconsin Press, Madison.

Webb, S. (1976). The volcanic envelope of the Coastal Batholith in Lima and Ancash, Peru, *Thesis Ph.D., University of Liverpool,* (unpubl.).

Wegmann, C.E., (1935). Zur Deutung der Migmatite, *Geol. Rundsch.*, 26, 305–350.

White, A.J.R. and Chappell, B.W. (1977). Ultrametamorphism and granitoid genesis, *Tectonophysics*, 43, 7–22.

White, A.J.R., Williams, I.S. and Chappell, B.W. (1977). Geology of the Berridale 1: 100,000 sheet (8625), *Mem. geol. Surv. New South Wales*, 138 pp.

Whitford, D.J. (1975). Strontium isotope studies of the volcanic rocks of the Saunda Arc. Indonesia, and their petrogenetic significance, *Geochim. cosmochim. Acta.*, 39, 1287–1302.

Whitford, D.J. and Bloomfield, K. (1975). Geochemistry of Late Cenozoic volcanic rocks from the Nevado de Toluca area, Mexico, *Carnegie Inst. of Washington Yearbook (1975/1976),* 207–213.

Wilson, J.F., Bickle, M.J., Hawkesworth, C.J., Martin, A., Nisbet, E. and Orpen, J.L. (1978). Granite – greenstone terrains of the Rhodesian Archaean craton, *Nature. London*, 271, 23–27.

Wilson, P.A. (1975). K-Ar age studies in Peru with special reference to the emplacement of the Coastal Batholith, *Thesis Ph.D., University of Liverpool* (unpubl.).

Windley, B.F. (1977). *The Evolving Continents*, Wiley, London, 385 pp.

147

Windley, B.F. and Smith, J.V. (1976). Archaean high grade complexes and modern continental margins, *Nature. London,* 260, 671–675.

Winkler. H. (1976). *Petrogenesis of metamorphic rocks.* 4th ed., 334 pp, Springer-Verlag.

Wood, D.A. (1978). Major and trace element variations in the Tertiary lavas of Eastern Iceland and their significance with respect to the Iceland geochemical anomaly. *J. Petrol.,* 19, 393–436.

Wood, D.A., Joron, J. -L., Marsh, N.G., Tarney, J. and Treuil, M. (in press). Major and trace element variations in basalts from the North Philippine Sea drilled by IPOD Leg 58: a comparative study of back-arc basin basalts with lava series from Japan and mid-ocean ridges, *Init. Repts. Deep Sea Drilling Project,* 58.

Wright, A.E. (1976). Alternating subduction directions and the evolution of the Atlantic Caledonides, *Nature. London,* 264, 156–160.

Wright, A.E., Tarney, J., Palmer, K.F., Moorlock, B.S.P. and Skinner, A.C. (1973). The geology of the Angmagssalik area, East Greenland and possible relationships with the Lewisian of Scotland. In: *The Early Precambrian of Scotland and related rocks of Greenland* (Eds. R.G. Park and J. Tarney) pp. 157–177; University of Keele.

Wyllie, P.J. (1977). Crustal anatexis: an experimental review. *Tectonophysics,* 43, 41–71.

Zambrano, J.J. and Urien, C.M. (1970). Geological outline of the basins in southern Argentina and their continuation off the Atlantic shore, *J. geophys. Res.,* 75, 1363–1396.

Zartman, R.E. (1974). Lead isotopic provinces in the Cordillera of the western United States and their geologic significance, *Econ. Geol.,* 69, 792–805.

Zwart, H.J. (1967). The duality of orogenic belts. *Geol. Mijnbouw,* 46, 283–309.

Zwart, H.J. (1969). Metamorphic facies series in the European orogenic belts and their bearing on the causes of orogeny, *Spec. Paper. Geol. Association. Can.,* 5, 7–16.